PLEASE

Engineers' Guide to Pressure Equipment

The Pocket Reference

Clifford Matthews
BSc, CEng, MBA, FIMechE

Professional Engineering Publishing

Professional Engineering Publishing Limited,
London and Bury St Edmunds, UK

First published 2001

This publication is copyright under the Berne Convention and the International Copyright Convention. All rights reserved. Apart from any fair dealing for the purpose of private study, research, criticism, or review, as permitted under the Copyright Designs and Patents Act 1988, no part may be reproduced, stored in a retrieval system, or transmitted in any form or by any means, electronic, electrical, chemical, mechanical, photocopying, recording or otherwise, without the prior permission of the copyright owners. Unlicensed multiple copying of this publication is illegal. Inquiries should be addressed to: The Publishing Editor, Professional Engineering Publishing Limited, Northgate Avenue, Bury St Edmunds, Suffolk, IP32 6BW, UK.

ISBN 1 86058 298 2

© 2001 Clifford Matthews

A CIP catalogue record for this book is available from the British Library.

This book is intended to assist engineers and designers in understanding and fulfilling their obligations and responsibilities. All interpretation contained in this publication – concerning technical, regulatory, and design information and data, unless specifically otherwise identified – carries no authority. The information given here is not intended to be used for the design, manufacture, repair, inspection, or certification of pressure equipment, whether or not that equipment is subject to design codes and statutory requirements. Engineers and designers dealing with pressure equipment should not use the information in this book to demonstrate compliance with any code, standard, or regulatory requirement. While great care has been taken in the preparation of this publication, neither the author nor the Publishers do warrant, guarantee, or make any representation regarding the use of this publication in terms of correctness, accuracy, reliability, currentness, comprehensiveness, or otherwise. Neither the Publisher, Author, nor anyone, nor anybody who has been involved in the creation, production, or delivery of this product shall be liable for any direct, indirect, consequential, or incidental damages arising from its use.

Printed and bound in Great Britain by St Edmundsbury Press Limited, Suffolk, UK

About the Author

Cliff has extensive experience as consulting/inspection engineer on power/chemical plant projects worldwide: Europe, Asia, Middle East, USA, Central and South America, and Africa. He has been an expert witness in a wide variety of insurance investigations and technical disputes in power plants, ships, paper mills, and glass plants concerning values of $40m. Cliff also performs factory inspections in all parts of the world including China, USA, Western and Eastern Europe. He carries out site engineering in the Caribbean – Jamaica, Bahamas, and the Cayman Islands.

Cliff is also the author of several books and training courses on pressure equipment-related related subjects.

Related Titles

IMechE Engineers' Data Book – Second Edition	C Matthews	1 86058 248 6
A Practical Guide to Engineering Failure Investigation	C Matthews	1 86058 086 6
Handbook of Mechanical Works Inspection – A guide to effective practice	C Matthews	1 86058 047 5
Process Machinery – Safety and Reliability	Edited by W Wong	1 86058 046 7
Plant Monitoring and Maintenance	IMechE Seminar	1 86058 087 4
Journal of Process Mechanical Engineering	Procedings of The IMechE, Part E	ISSN 0954-4089

For the full range of titles published, by Professional Engineering Publishing contact:
Sales Department
Professional Engineering Publishing Limited
Northgate Avenue
Bury St Edmunds
Suffolk IP32 6BW
UK
Tel: +44 (0) 1284 724384; Fax: +44 (0) 1284 718692
E-mail: sales@pepublishing.com
www. pepublishing.com

Contents

Foreword	*xi*
Chapter 1 Websites: Quick Reference	**1**
1.1 Organizations and associations	1
1.2 General technical information	4
1.3 Directives and legislation	6
1.4 The KKS power plant classification system	7
Chapter 2 Pressure Equipment Types and Components	**11**
2.1 What is pressure equipment?	11
2.2 Pressure equipment categories	11
2.3 Pressure equipment symbols	13
Chapter 3 Basic Design	**21**
3.1 Introduction – the influence of codes and standards	21
3.2 Vessel design – basic points	21
3.2.1 Design basis	21
3.2.2 Safety first – corrosion allowance and welded joint efficiency	23
3.2.3 Pressure vessel cylinders	24
3.2.4 Vessel classes	25
3.2.5 Heads	25
3.2.6 Openings and compensation	29
3.2.7 Inspection openings	32
3.2.8 Pipes and flanges	33
3.2.9 Pads	33
3.2.10 Vessel supports	34
3.3 Simple pressure vessels (SPVs) – basic design	35
3.3.1 Material selection	35
3.3.2 Welds	37
3.3.3 Stress calculations	37

3.4	Gas cylinders – basic design	39
3.5	Heat exchangers – basic design	47
	3.5.1 Contact-type exchangers	47
	3.5.2 Surface-type exchangers	47
	3.5.3 Thermal design	47
	3.5.4 Special applications	53
3.6	Design by Analysis (DBA) – pr EN 13445	53
	3.6.1 What does DBA offer?	53
	3.6.2 How does DBA fit into pr EN 13445?	55
	3.6.3 DBA – the technical basis	55

Chapter 4 Applications of Pressure Vessel Codes — 59

4.1	Principles	59
4.2	Code compliance and intent	59
4.3	Inspection and test plans (ITPs)	60
4.4	Important code content	63
4.5	PD 5500	64
	4.5.1 PD 5500 and the PED ESRs	70
4.6	The ASME vessel codes	76
	4.6.1 Summary	76
	4.6.2 Allowable stresses	81
	4.6.3 Cylindrical vessel shells	81
	4.6.4 Flat plates, covers, and flanges	85
	4.6.5 Vessel openings – general	90
	4.6.6 Heat exchangers	91
	4.6.7 Special analyses	91
	4.6.8 ASME 'intent'	95
4.7	TRD	96
4.8	Air receivers	98
4.9	Shell boilers: BS 2790 and EN 12953	101
4.10	Canadian standards association B51-97, part 1 boiler, pressure vessel, and piping code – 1997	106
4.11	CODAP – unfired pressure vessels	107
4.12	Water tube boilers: BS 1113/pr EN 12952	107
4.13	Materials and referenced standards – quick reference	109
4.14	Pressure vessel codes – some referenced standards	111

Chapter 5 Manufacture, QA, Inspection, and Testing — 113

5.1	Manufacturing methods and processes	113
5.2	Vessel visual and dimensional examinations	114
	5.2.1 The vessel visual examination	114

		5.2.2	The vessel dimensional check	116
		5.2.3	Vessel markings	118
5.3	Misalignment and distortion			118
		5.3.1	What causes misalignment and distortion?	118
		5.3.2	Toleranced features	119
5.4	Pressure and leak testing			122
		5.4.1	The point of a pressure test	122
		5.4.2	The standard hydrostatic test	123
		5.4.3	Pneumatic testing	124
		5.4.4	Vacuum leak testing	125
5.5	ASME certification			126
		5.5.1	The role of the AI (Authorized Inspector)	126
		5.5.2	Manufacturers' data report forms	127
		5.5.3	The code symbol stamps	129
		5.5.4	ASME and the European Pressure Equipment Directive (PED)	131
5.6	European inspection terms and bodies: EN 45004: 1995			132
5.7	The role of ISO 9000			133
		5.7.1	The objectives of the changes	133
		5.7.2	What will the new standards be?	134
		5.7.3	What are the implications?	134
		5.7.4	The 'new format' ISO 9001: 2000	134

Chapter 6 Flanges, Nozzles, Valves, and Fittings — 137

6.1	Flanges			137
6.2	Valves			141
		6.2.1	Types of valves	141
		6.2.2	Valve technical standards	141
6.3	Safety devices			151
		6.3.1	Safety relief valves – principles of operation	152
		6.3.2	Terminology – safety valves	153
6.4	Nozzles			155
6.5	Power piping – ASME/ANSI B31.1 code			158
6.6	Fittings			161
		6.6.1	Pressure equipment fittings	161
		6.6.2	Pipework classification	161

Chapter 7 Boilers and HRSGs — 167

7.1	Fundamentals of heat transfer			167
		7.1.1	Specific heat, c	167
		7.1.2	Enthalpy, h	167

	7.1.3	Latent heat	168
	7.1.4	Steam characteristics	168
	7.1.5	Gas characteristics	173
7.2	Heat recovery steam generators (HRSGs)		173
	7.2.1	General description	173
	7.2.2	HRSG operation	176
	7.2.3	HRSG terms and definitions	180
	7.2.4	HRSG materials	183

Chapter 8 Materials of Construction — 185
8.1	Plain carbon steels — basic data	185
8.2	Alloy steels	185
8.3	Stainless steels – basic data	186
8.4	Non-ferrous alloys – basic data	189
8.5	Material traceability	190
8.6	Materials standards – references	192

Chapter 9 Welding and NDT — 195
9.1	Weld types and symbols		195
9.2	Weld processes		195
9.3	Welding standards and procedures		203
9.4	Destructive testing of welds		205
	9.4.1	Test plates	205
	9.4.2	The tests	205
9.5	Non-destructive testing (NDT) techniques		209
	9.5.1	Visual examination	209
	9.5.2	Dye penetrant (DP) testing	209
	9.5.3	Magnetic particle (MP) testing	212
	9.5.4	Ultrasonic testing (UT)	213
	9.5.5	Radiographic testing (RT)	219
9.6	NDT acronyms		223
9.7	NDT: vessel code applications		225
9.8	NDT standards and references		227

Chapter 10 Failure — 229
10.1	How pressure equipment materials fail		229
	10.1.1	LEFM method	230
	10.1.2	Multi-axis stresses states	231
10.2	Fatigue		232
	10.2.1	Typical pressure equipment material fatigue limits	233
	10.2.2	Fatigue strength – rules of thumb	234

10.3	Creep		235
10.4	Corrosion		238
	10.4.1	Types of corrosion	238
	10.4.2	Useful references	241
10.5	Boiler failure modes		241
10.6	Failure-related terminology		244

Chapter 11 Pressure Equipment: Directives and Legislation — 249

11.1	Introduction: what's this all about?		249
	11.1.1	The driving forces	249
	11.1.2	The EU 'new approaches'	250
11.2	The role of technical standards		250
	11.2.1	Harmonized standards	250
	11.2.2	National standards	251
	11.2.3	The situation for pressure equipment	251
11.3	Vessel 'statutory' certification		253
	11.3.1	Why was certification needed?	253
	11.3.2	What was certification?	253
	11.3.3	Who could certificate vessels?	254
11.4	The CE mark – what is it?		255
11.5	Simple pressure vessels		255
11.6	The simple pressure vessels directive and regulations		256
	11.6.1	SPVs – summary	256
	11.6.2	Categories of SPVs	257
	11.6.3	SPV harmonized standards	264
11.7	Transportable pressure receptacles: legislation and regulations		265
	11.7.1	TPRs legislation	265
11.8	The pressure equipment directive (PED) 97/23/EC		271
	11.8.1	PED summary	271
	11.8.2	PED – its purpose	273
	11.8.3	PED – its scope	273
	11.8.4	PED – its structure	274
	11.8.5	PED – conformity assessment procedures	275
	11.8.6	Essential safety requirements (ESRs)	294
	11.8.7	Declaration of conformity	311
	11.8.8	Pressure equipment marking	312
11.9	Pressure Equipment Regulations 1999		312
	11.9.1	The Pressure Equipment regulations – structure	312
11.10	Notified Bodies		314
	11.10.1	What are they?	314

	11.10.2	UK Notified Bodies	314
11.11	Sources of information		317
	11.11.1	Pressure system safety – general	317
	11.11.2	Transportable pressure receptacles (gas cylinders)	318
	11.11.3	The simple pressure vessel directive/regulations	318
	11.11.4	The pressure equipment directive	318
	11.11.5	The pressure equipment regulations	319
	11.11.6	PSSRs and written schemes	319

Chapter 12 In-service Inspection — 321
12.1 A bit of history — 321
12.2 The Pressure Systems Safety Regulations (PSSRs) 2000 — 322

Chapter 13 References and Information Sources — 325
13.1 European Pressure Equipment Research Council (EPERC) — 325
13.2 European and American associations and organizations relevant to pressure equipment activities — 327
13.3 Pressure vessel technology references — 335

Appendix 1 Steam Properties Data — 337

Appendix 2 Some European Notified Bodies (PED) — 343
Notified Bodies (PED Article 12) — 343
Recognized Third-Party Organizations (PED Article 13) — 348

Appendix 3 Standards and Directives Current Status — 351

Index — 383

Foreword

During the course of my career working with pressure equipment, I have come to realize the value of concise, introductory information such as is provided in *Engineers' Guide to Pressure Equipment*. It provides practising engineers with a unique collection of essential, up-to-date information, in a practical and easy-to-use form. This guide is most welcome within the engineering community.

Pressure equipment technology continues to expand rapidly presenting ever more complex and challenging issues and problems to the engineer, safety expert, or operative. Recent developments in international, particularly European, legislation, technical codes, and standards are resulting in exciting changes within the industry. This handy reference – Engineers' Guide to Pressure Equipment, has been compiled in co-operation with several highly experienced experts in the pressure equipment field. It is my hope that it will be the first of many editions that will serve as a ready source of data and information on legislative and technical developments.

This pocket guide provides an essential introduction to the technical and administrative aspects of vessel manufacture and use. Engineers dealing with rotating machinery and those concerned with fabrications and pressure vessel manufacture will find this book of immense value and interest.

Clifford Matthews has produced a work which fulfils a major role in the industry and is set to become a permanent feature.

Mr N Haver BEng
Senior Pressure Systems Certification Engineer

CHAPTER 1

Websites: Quick Reference

1.1 Organizations and associations

Organization	URL
AEA Technology plc	www.aeat.co.uk
American Bureau of Shipping	www.eagle.org
American Consulting Engineers Council	www.acec.org
American Institute of Engineers	www.members-aie.org
American Institute of Steel Construction Inc.	www.aisc.org
American Iron and Steel Institute	www.steel.org
American National Standards Institute	www.ansi.org
American Nuclear Society	www.ans.org
American Petroleum Institute	www.api.org
American Society for Non-Destructive Testing	www.asnt.org
American Society for Testing of Materials	www.ansi.org
American Society of Heating, Refrigeration and Air Conditioning Engineers	www.ashrae.org
American Society of Mechanical Engineers	www.asme.org
American Water Works Association Inc.	www.awwa.org
American Welding Society	www.awweld.org
APAVE Limited	www.apave-uk.com
Association of Iron and Steel Engineers (USA)	www.aise.org

British Inspecting Engineers	www.bie-international.com
British Institute of Non-Destructive Testing	www.bindt.org
British Standards Institution	www.bsi.org.uk
British Valve and Actuator Manufacturers Association	www.bvama.org.uk
Det Norske Veritas	www.dnv.com
DTI STRD 5:	www.dti.gov.uk/strd
Engineering Integrity Society	www.demon.co.uk/e-i-s
European Committee for Standardisation	www.cenorm.be www. newapproach.org
Factory Mutual Global (USA)	www.fmglobal.com
Fluid Controls Institute Inc. (USA)	www.fluidcontrolsinstitute.org
Hartford Steam Boiler (USA)	www.hsb.com
Heat Transfer Research Inc. (USA)	www.htrinet.com
Her Majesty's Stationery Office	www.hmso.gov.uk/legis.htm www.hmso.gov.uk/si www.legislation.hmso.gov.uk
HSB Inspection Quality Limited	www.hsbiql.co.uk
HSE (Home page)	www.hse.gov.uk/hsehome.htm
International Standards Organization	www.iso.ch
Lloyd's Register	www.lrqa.com
Manufacturers Standardization Society of the Valve and Fittings Industry (USA)	www.mss-hq.com
National Association of Corrosion Engineers (USA)	www.nace.org
National Board of Boiler and Pressure Vessel Inspectors (USA)	www.nationalboard.org
National Fire Protection Association (USA)	www.nfpa.org
National Fluid Power Association (USA)	www.nfpa.com
National Institute of Standards and Technology (USA)	www.nist.gov

Pipe Fabrication Institute (USA)	www.pfi-institute.org
Plant Safety Limited	www.plantsafety.co.uk
Registrar Accreditation Board (USA)	www.rabnet.com
Royal Sun Alliance Certification Services Limited	www.royal-and-sunalliance.com/
Safety Assessment Federation	www.safed.co.uk
SGS (UK) Limited	www.sgs.com
The Aluminum Association Inc. (USA)	www.aluminum.org
The Engineering Council	www.engc.org.uk
The Institute of Corrosion	www.icorr.demon.co.uk
The Institute of Energy	www.instenergy.org.uk
The Institute of Materials	www.instmat.co.uk
The Institute of Quality Assurance	www.iqa.org
The Institution of Mechanical Engineers (IMechE) Pressure Systems Group	www.imeche.org.uk www.imeche.org.uk/pressure/index.htm
The Institution of Plant Engineers	www.iplante.org.uk
The United Kingdom Accreditation Service	www.ukas.com
The Welding and Joining Society	www.twi.co.uk/members.wjsinfo.html
Tubular Exchanger Manufacturers Association Inc. (USA)	www.tema.org
TUV(UK) Limited	www.tuv-uk.com
TWI Certification Limited	www.twi.co.uk
Zurich Engineering Limited	www.zuricheng.co.uk

1.2 General technical information

Technical standards	www.icrank.com/Specsearch.htm www.nssn.org
Pressure vessel design	www.birdsoft.demon.co.uk/englib/pvessel www.normas.com/ASME/BPVC/guide.html www.pretex.com/Glossary.html www.nationalboard.org/Codes/asme-x.html
Boilers-fire tube	www.kewaneeintl.com/scotch/scotivOb.ht www.hotbot.com/books/vganapathy/firewat.html
Pressure vessel software	www.chempute.com/pressure_vessel.htm www.eperc.jrc.nl www.mecheng.asme.org/ www.coade.com/pcodec/c.htm www.codeware.com/ www.ohmtech.no/
Transportable pressure receptacles (gas cylinders)	www.hse.gov.uk/spd/spdtpr.htm www.iso.ch/cate/2302030.html www.hmso.gov.uk/sr/sr1998/19980438.htm www.pp.okstate.edu/ehs/links/gas.htm www.healthandsafety.co.uk/E00800.html www.hmso.gov.uk/si/si1998/19982885.htm
Valves	www.yahoo.com/Business_and_Economy/Companies/Industrial/Valves_and_Control www.highpressure.com www.flowbiz.com/valve_manual.htm www.valvesinternational.co.za/further.htm
Heat exchangers	www.britannica.com/seo/h/heat-exchanger/ www.geapcs.com www.chem.eng.usyd.edu.au/pgrad/bruce/h www.firstworldwide.com/heatex.htm www.heat-exchangers.com/heat-exchanger
Simple pressure vessels	www.egadvies.nl/ce/scope/DrukvateEN.html www.dti.gov.uk/strd
Boilers and HRSGs	www.normas.com/ASME/BPVC/qx0010.html www.boiler-s.com/boiler-spages.hotbot.com/books/vganapathy/boilers.html www.abb.com/americas/usa/hrsg.htm www.hrsg.com/gatorpwr.che.ufl.edu/cogen/equipment/codes/PTC4.4/Default.asp.

Welding	www.welding.com/
	www.cybcon.com/Nthelen/1weld.html
	www.amweld.org/
	www.mech.uwa.edu.au/DANotes/
	welds/home.html
Non-destructive testing	www.ndt.net/
	www.dynatup.com/apps/glossary/C.htm
Failure	www.sandia.gov/eqrc/e7rinfo.html
	www.netaccess.on.ca/~dbc/cic_hamilton/
	krissol.kriss.re.kr/failure/STRUC/MATE
	www.clihouston.com/asmfailureanalysis.htm
Corrosion	www.cp.umist.ac.uk/
	www.icorr.demon.co.uk/about.html
	www.cranfield.ac.uk/cils/library/subje
Technical reference books	www.icrank.com/books.htm
	www.powells.com/psection/MechanicalEng
	www.lib.uwaterloo.ca/discipline/mechen
	www.engineering-software.com/
	www.fullnet.net/dbgnum/pressure.htm
	www.engineers4engineers.co.uk/0938-but
ISO 9000	www.isoeasy.org/
	www.startfm3.html
Forged flanges	www.kotis.net/~porls2/ForgedFL.htm
	www.englink.co.za/htmlfiles/nclro2/BS3.1.htm
	www.maintenanceresources.com/Bookstore
	polyhydron.com/hhpl/SAI.htm
Creep	www.men.bris.ac.uk/research/material/
	projects/fad.htm/
Air receivers	www.manchestertank.com/hor400.htm
Materials	www.Matweb.com
	www.pump.net/otherdata/pdcarbonalloysteel.htm
General on-line reference websites	www.efunda.com/home.cfm
	www.flinthills.com/~ramsdale/EngZone/d
	www.eevl.ac.uk
Safety valves	www.taylorvalve.com/safetyreliefvalves.htm
Control valves	www.halliburton.com/deg/dvd/masoneilan/
	80000_prod.asp?print=yes

1.3 Directives and legislation

For a list of new EC directives and standards on pressure equipment, go to www.nssn.org and search using keywords 'pressure vessel standards' and 'EC'.

- The entry website for the **Health and Safety Executive** (HSE) is:
  www.hse.gov.uk/hsehome.htm
- **Statutory Instrument** (SI) documents are available from Her Majesty's Stationery Office (HMSO) at:
  www.hmso.gov.uk/legis.htm
  www.hmso.gov.uk/si
  www.legislation.hmso.gov.uk
- The **Department of Trade and Industry** (DTI) entry web site is:
  www.dti.gov.uk/strd
- Background information on mutual recognition agreements relating to **European directives** in general is available on:
  http://europa.eu.int/comm/enterprise/international/indexb1/htm
- Reference to most **pressure equipment-related directives** (and their interpretation guidelines) is available on:
  www.tukes.fi/english/pressure/directives_and_guidelines/index.htm
- A good general introduction to the **Pressure Equipment Directive** (PED) is available on:
  www.ped.eurodyne.com/directive/directive.html
  www.dti.gov.uk/strd/pressure.htm
- **Guides on the PED** from the UK DTI are available for download on:
  www.dti.gov.uk/strd/strdpubs.htm
- The **European Commission Pressure Equipment Directive** website has more detailed information on:
  europa.eu.int/comm/dg03/directs/dg3d/d2/presves/preseq.htm
  or
  europa.eu.int/comm/dg03/directs/dg3d/d2/presves/preseq1.htm
- The **CEN** website provides details of all harmonized standards:
  www.newapproach.org/directivelist.asp
- For details of **harmonized standards** published in the EU official journal, go to:
  www.europa.eu.int/comm/enterprise/newapproach/standardization/harmstds/reflist.html

- A listing of **European Notified Bodies** for the PED is available from:
  www.conformance.co.uk/CE_MARKING/ce_notified.html
- All the sections of the text of the **UK Pressure Equipment Regulations** are available on:
  http://www.hmso.gov.uk/si/si1999/19992001.htm
- The **UK Pressure Systems Safety Regulation** (PSSRs) are available from:
  www.hmso.gov.uksi/si2000/20000128.htm

1.4 The KKS power plant classification system

The KKS (Kraftwerk Kennzeichensystem) is a completely generic designation system that identifies every system, component, and location within a plant, using a string of letters and numbers. It is used for power and process plant across many industry sectors. The KKS designations are commonly used on equipment/spares inventories, Process and Instrumentation Diagrams (PIDs), and plant labelling. From the pressure equipment viewpoint, the most important function is that related to process designation.

The KKS process classification consists of a maximum of four levels: 0, 1, 2, and 3 (see Fig. 1.1). Level 0 identifies the overall plant and differentiates between units or blocks. Level 1 identifies the plant system, classifying each area with a three-letter code – the first letter specifying the function group, and two further letters. Level 2 identifies individual systems and equipment items within the function groups, with level 3 available for further sub-designation. Table 1.1 shows a small sample of the thousands of codes available.

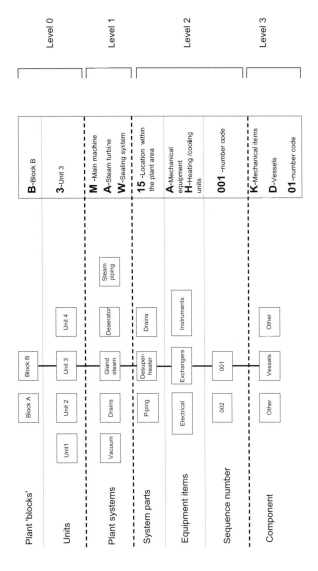

Gland Steam Desuperheater
B3MAW15AH001KD 01

Fig. 1.1 Structure of the KKS plant classification system

Table 1.1 Typical KKS codes

Level 1

First letter	Identifies	Second letter	Identifies	Third letter	Identifies
C	Control equipment	MA	Steam turbine	MAA	HP turbine
G	Water system	MB	Gas turbine	MAB	IP turbine
H	Heat generation	ME	Water turbine	MAC	LP turbine
L	Steam and gas cycles	MM	Compressor	MAG	Condenser
M	Main machines	MV	Oil system	MAJ	Air system
P	Cooling water			MAL	Drains
R	Gas generation			MAM	Leak-off system
				MAQ	Vents

Level 2

First letter	Identifies	Second letter	Identifies
A	Mechanical equipment	AA	Valves
B	Auxiliary equipment	AB	Isolating elements
C	Measuring circuits	AC	Heat exchangers
D	Control circuits	AH	Heating units
G	Electrical equipment	AN	Compressors
H	Machinery sub-assembly	AP	Pumps
		AV	Combustion units
		BB	Tanks
		BN	Ejectors
		BR	Piping

Level 3

First letter	Identifies	Second letter	Identifies
K	Mechanical items	KA	Valves
M	Mechanical items	KC	Heat exchangers
		KD	Vessels/tanks
		KN	Compressors
		KP	Pumps
		MR	Piping
		MT	Turbines

CHAPTER 2

Pressure Equipment Types and Components

2.1 What is pressure equipment?

A unanimously agreed definition of what constitutes 'pressure equipment' is hard to find. The terms *pressure equipment*, *pressure vessel*, and *pressure system*, while appearing technically straightforward, are bound up as the subject matter of discussion by the numerous technical committees that draft legislation, regulations, and technical standards. Due to the way that various international and national directives, standards, etc., are structured, there are often differences and contradictions in meaning between technical terms. The end result is that definition of what does, or does not, constitute pressure equipment may differ between countries, industry, and technical application. It is a complex and ever-changing picture.

Thankfully, fundamental technical aspects of pressure equipment do not change that quickly. Design and manufacturing practices have been developed over the past 100 years and the technology of pressure equipment is well supported by technical standards and codes of practice from many of the world's developed countries.

2.2 Pressure equipment categories

Table 2.1 shows the scope of some common technical categories of pressure equipment used in the mainstream engineering industries. Note that these categories represent arbitrary technical divisions only and have no direct relevance to the inclusion of the equipment under any directives or regulations.

From the engineering viewpoint, pressure equipment types, and components are all surprisingly similar. Their characteristics are outlined in Table 2.2.

Table 2.1 Pressure equipment categories*

Simple pressure vessels	Gas cylinders	Unfired pressure vessels	Boilers	Valves	Pipework	Miscellaneous
Simple receivers (unfired)	**LPG cylinders (transportable)**	Complex air receivers	**Heating boilers**	**Stop valves**	Pipelines	Turbines
Air	Two-piece cylinders	Boiler drums	Package 'shell' boilers	Plug	Power piping	Autoclaves
Nitrogen	Three-piece cylinders	Pressurized storage containers	**Coal/oil-fired power boilers**	Globe	Service piping	Road transport vehicles
Low-pressure/ atmospheric tanks		Condensers	**General water tube boilers**	Gate	Flanges	Road and rail vehicle components (brakes, etc.)
		Liquid cooled		Butterfly	Pipework fittings	
		Air cooled	**HRSGs**	**Check valves**	Pressurized accessories	Portable tools
		Superheaters	Fired HRSGs	Stop check		Domestic heating systems
		Desuperheaters	Unfired HRSGs	Lift check		
		Economizers	Forced circulation HRSGs	Tilting disc check		Aircraft components
		Heat exchangers	Natural circulation HRSGs	Swing check		Gas-loaded hydraulic accumulators
		High-pressure feed heaters	Single-pressure HRSGs	**Safety valves**		Vapour compression refrigeration systems
		Tube type exchangers	Multiple-pressure HRSGs	Pressure relief		
		Plate type exchangers		Vacuum relief		Engine cooling systems
		Contact exchangers		**Control valves**		
		Chemical process/reaction vessels		Two-way valves		
		Metal vessels		Three-way valves		
		GRP vessels		Regulating valves		
		Nuclear vessels		**Metering valves**		
				Needle valves		
				Diaphragm valves		

* Note that these only show broad technical categories of pressure equipment. This table does not infer the applicability of any directives, regulations, or technical standards – this is covered in Chapter 11.

Table 2.2 Some characteristics of 'pressure equipment'

Gauge pressure	Equipment is subject to a positive gauge pressure or negative gauge (vacuum) pressure.
Principal stresses	Components are subject to principal stresses in three dimensions, or two-dimensional membrane stresses (for thin-walled shells).
Stored energy	Pressure equipment, almost by definition, acts to contain stored energy in use. Such stored energy can constitute a hazard.
Controlled manufacture	Due to the potential hazard if failure occurs, pressure equipment is subject to controls on its specification, design, and manufacture. The amount of control varies, depending on what the equipment is and how it will be used.
Factors of safety	All pressure equipment has factors of safety incorporated into its design; these provide a margin against unforeseen circumstances and reduce the risk of failure to acceptable levels.
Inspection and testing	These play an important part in ensuring the fitness for purpose of pressure equipment during its manufacture and before use.
In-service inspection	As a general rule, it is necessary to inspect pressure equipment throughout its working life to make sure it continues to be safe and fit for purpose.

2.3 Pressure equipment symbols

Pressure equipment symbols are in regular use in schematic drawings and Process and Instrumentation Diagrams (PIDs). There are many variations based on BS, ISO, and American standards. Figures 2.1–2.4 show some commonly used types. Table 2.3 and Figs 2.5 and 2.6 show symbols for typical pressure equipment-related instrumentation.

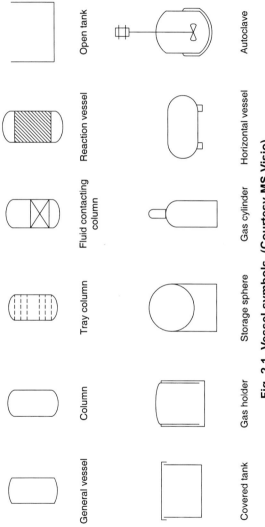

Fig. 2.1 Vessel symbols. (Courtesy MS Visio)

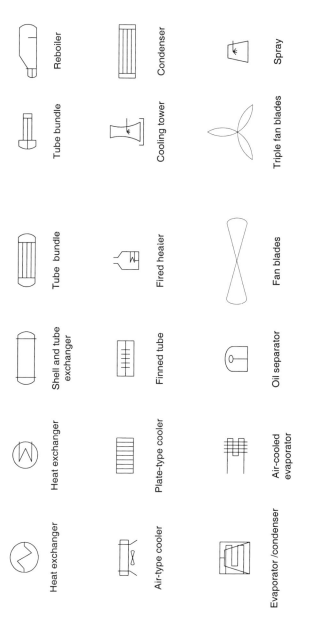

Fig. 2.2 Heat exchanger symbols. (Courtesy MS Visio)

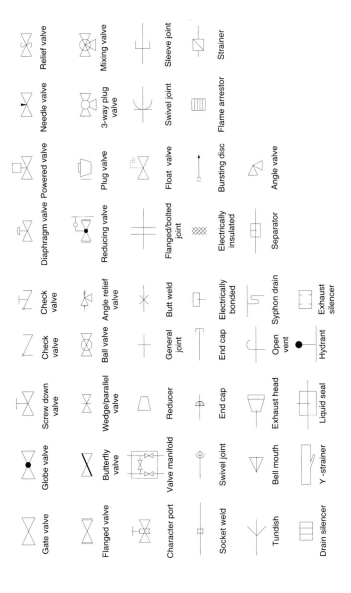

Fig. 2.3 Valves and fittings. (Courtesy MS Visio)

Fig. 2.4 Piping and fittings symbols. (Courtesy MS Visio)

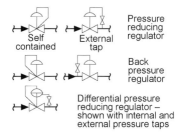

Self-actuated regulators

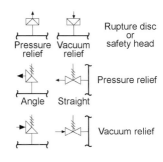

Relief valves

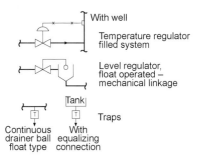

Temperature and level regulators

Fig. 2.5 Valve/regulator symbols from ISA S5.1

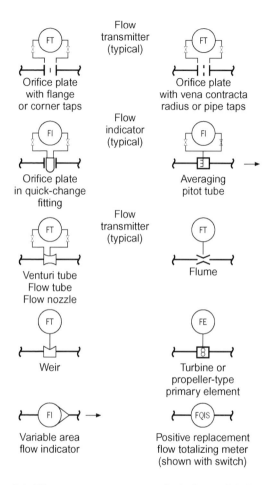

Fig. 2.6 Flow measurement symbols from ISA S5.1

Table 2.3 Typical instrument identification letters

	First letters Measurement	Modifier	*Succeeding letters* Readout or passive function	Output function	Modifier
A =	Analysis		Alarm		
B =	Burner, combustion		User's choice	User's choice	User's choice
C =	Conductivity (electrical)	Control			
D =	Density or Sp. Gr		Differential		
E =	Voltage (Emf)		Sensor (primary element)		
F =	Flow rate	Ratio (fraction)			
G =	User's choice		Glass or viewing device		
H =	Hand (manual)				High
I =	Current (electrical)		Indicate		
J =	Power	Scan			
K =	Time or time schedule	Time rate of change		Control station	
L =	Level		Light		Low
M =	User's choice	Momentary			Middle or intermediate
N =	User's choice		User's choice	User's choice	User's choice
O =	User's choice		Office (restriction)		
P =	Pressure or vacuum		Test or sample point		
Q =	Quantity	Integrate or totalize			
R =	Radiation		Record or print		
S =	Speed or frequency	Safety		Switch	
T =	Temperature			Transmit	
U =	Multi-variable		Multi-function	Multi-function	Multi-function
V =	Vibration. mech. analysis			Valve. Damper or louvre	
W =	Weight or force		Well		
X =	Unclassified	X axis	Unclassified	Unclassified	Unclassified
Y =	Event state or presence	Y axis		Relay. Compute. Convert	
Z =	Position or dimension	Z axis		Drive. Actuator. Etc.	

CHAPTER 3

Basic Design

3.1 Introduction – the influence of codes and standards

Published codes and standards have a guiding role in the design and manufacture of pressure equipment. In contrast to many other types of engineering equipment, codes and standards for pressure equipment provide comprehensive coverage of all aspects, from design and manufacture, through to works inspection, testing, and in-use inspection.

The design of equipment features heavily in pressure equipment technical standards. Engineering interpretation of the mechanics of stress, fatigue, creep, and other mechanisms is given in great detail. Ideally, all technical standards would use the same technical assumptions and, therefore, come to the same design conclusions. The practical situation is not so straightforward; standards from different countries and technical bodies occasionally manage to reach different conclusions about the same thing. Expect, therefore, to see some technical standards taking a different approach, and reaching different technical conclusions, to others. This does not necessarily mean that a particular technical standard is right or wrong – only that it may be *different* to some of the others.

3.2 Vessel design – basic points

3.2.1 Design basis

Most pressure vessels are cylindrical in form and are designed using cylindrical shell theory. There are various practical requirements such as the need for cylinder end closures, holes for inlet/outlet pipes, and attachments. Further design criteria must then be applied to take account of the

probability of weld flaws and similar defects. Figure 3.1 shows a typical 'generic' representation of a pressure vessel. Basic vessel design is concerned with internally pressurized, welded steel, unfired vessels operating at room temperature and above, thereby avoiding the complexities associated with:
- buckling due to external pressure and manufacturing inaccuracies;
- the effects of low temperatures on material properties.

A basic feature of all vessel design codes is the limitation placed on design tensile stress, S, of the vessel material. A basic methodology is to define acceptable design tensile stress as the minimum of:

$(S_y/A, \; S_u/B, \; S_{yT}/C, \; S_{crT}/D)$

where
$\quad S_y \quad$ = tensile yield strength (R_e) at room temperature
$\quad S_u \quad$ = ultimate tensile strength (R_m) at room temperature
$\quad S_{yT} \quad$ = tensile yield strength at design temperature
$\quad S_{crT} \quad$ = 100 000 h creep rupture strength at design temperature

A, B, C, and D are numerical factors that differ between design codes. Note also how a variety of symbols are used – European and US codes use different symbol sets, and both types are in common use. In general, US practice is to represent stress by the symbol S (or F), while in Europe, the use of σ is more common.

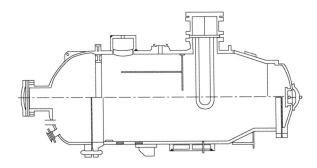

Fig. 3.1 A generic representation of a pressure vessel

3.2.2 Safety first – corrosion allowance and welded joint efficiency

Corrosion allowance

Corrosion occurring over the life of a vessel is catered for by a corrosion allowance, c, the design value of which depends upon the vessel duty and the corrosiveness of its contents. As an example, a design criterion of $c = 1$ mm is typical for air receivers in which condensation of air moisture is expected. Most vessel codes use the principle that, when dimensions in any formula refer to a corrodable surface, then the dimensions inserted into the formula are those at the *end* of the vessel's life, i.e. when all the corrosion allowance has been used up. Therefore, if a plate of nominal thickness, T, is subject to corrosion on one side, then $T - c$ must be substituted whenever nominal thickness appears in an equation. Similarly, if a tube of bore D_i corrodes, then at the end of its life the bore diameter will be $D_i + 2c$.

Welded joint efficiency

Most vessel codes assume that welded joints are not as strong as the parent plate, unless they are exhaustively inspected during manufacture and repaired if defects are found. This strength reduction is characterized by the *welded joint efficiency*.

> Welded joint efficiency, η = joint strength/parent material strength. It varies from 100 per cent for a perfect weld down to 75–85 per cent for welds in which integrity is not so assured.

Figure 3.2 shows the principle. The three bars (i, ii, iii) are all of the same width b and material strength S; they are loaded by the tensile load P. Bar (i) is seamless with a thickness t_s. Bar (ii) is welded with joint efficiency η, and thickness t_w locally at the joint. The safety factor in bar (ii) is Sbt_s/P away from the joint, and $(\eta S) bt_w /P$ at the joint. If these safety factors are designed to be identical (i.e. the bar is not to be weakened by the joint) then t_w must equal t_s/η. The economics of vessel manufacture generally do not favour an increase of bar thickness locally at a joint, so practically the whole bar would be designed with material thickness t_w to cater for an isolated joint (bar iii).

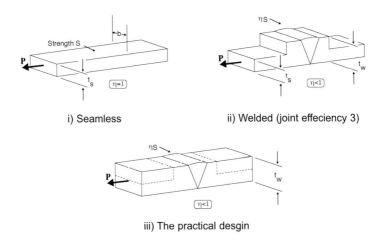

Fig. 3.2 Welded joint efficiency – the principle

3.2.3 *Pressure vessel cylinders*

Vessel cylinders are usually made from flat plates that are rolled and then welded along their longitudinal joints. Circumferential joints are used to attach end pieces (dished ends or 'heads') to the cylinder, and to weld together rolled plates for a long vessel. Weld types and efficiencies usually differ for longitudinal and circumferential joints. The joint stresses in a vessel must, therefore, be designed to satisfy both requirements.

Typically, this is expressed in a form something like (see Fig. 3.3).
- Circumferential stress, $S_c = (pD/2t) \leq \eta_\ell S$

where η_ℓ = longitudinal joint efficiency and p = pressure
and
- Longitudinal stress, $S_\ell = (pD/4t) \leq \eta_c S$

where η_c = circumferential joint efficiency and p = pressure
The symbol σ_c is sometimes used instead of S_c, and σ_ℓ instead of S_ℓ

p = pressure

In these formulae, most vessel codes take the diameter, D, to be the mean at the wall mid-surface ($D = D_i + t$), although they often quote the equation conveniently in terms of D_i (the inner diameter). As the circumferential stress is twice the longitudinal stress, the circumferential strength is usually the controlling parameter, provided that η_c is greater than half η_ℓ (which is

Basic Design

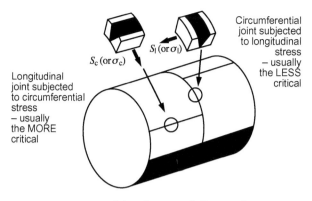

Fig. 3.3 Longitudinal and circumferential vessel stresses

normally the case). A plate thickness T, such that $(T - c) \geq t$, would be chosen from a commercially available range.

3.2.4 Vessel classes

Most vessel codes define several (typically 2 to 4) classes of vessel manufacture. The major differences between the classes vary from code to code (see code summaries given in Chapter 4).

3.2.5 Heads

Vessel heads are designed based on the theory of 'thin shells of revolution'. The basis is that the shell is formed conceptually by rotating the meridian, a curved line of selected shape lying in the r–z meridional plane, about the z axis (see Fig. 3.4). The resulting surface of revolution is clothed by a small, symmetrically disposed thickness, t, and the resulting shell loaded by internal pressure, p. The resulting shell is similar to a thin cylinder in that radial stresses are negligible and the membrane stresses are:
- the circumferential or hoop stress S_q or σ_θ
- the meridional stress S_f or σ_ϕ

Both these stresses can be found from an 'equilibrium' consideration since they, and the loading, are axisymmetric. Figure 3.5 illustrates this situation.

Consider the element located at point A in the r–z plane as shown, and defined by ϕ, dϕ, and dθ. The local surface normally cuts the z axis at point B, AB being defined as the radius, r_θ. The centre of curvature lies at C on the

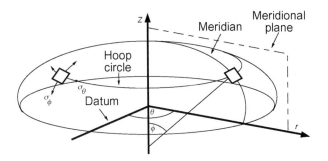

Fig. 3.4 Stresses in vessel heads

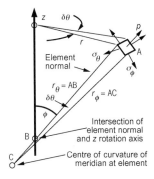

Fig. 3.5 Vessel head stresses – the derivation

normal, AC being the instantaneous radius of curvature of the meridian, r_ϕ.

The components of the pressure and stress resultants along the outward normal are:
- pressure: $p\, dA = p\, (r\, d\theta)\, r_\phi\, d\phi$;
- meridional stress: $-2\, S_\phi (t\, r d\theta) \sin (d\phi/2)$;
- circumferential stress: $-2\, S_\theta (t\, r_\phi d\phi) \sin (d\theta/2).\sin \phi$.

Taking limits and replacing $r = r_\theta \sin \phi$, then equilibrium of the element requires that

$$S_\theta / r_\theta + S_\phi / r_\phi = p / t \qquad \text{i.e. the so-called } \textit{membrane equation}$$

and for vertical equilibrium of the dish area above the hoop

$$pr^2\, p = 2\, p\, r\, t\, S_\phi \sin \phi$$

Solving these two equations gives the stress components in terms of r_θ and r_ϕ, which are in turn functions solely of the meridional geometry in terms of its shape and location with respect to the rotation axis

$$S_\phi = (p/2t)\, r_\theta \qquad S_\theta = S_\phi\,(2 - r_\theta/r_\phi)$$

Some typical specializations of this equation are outlined below.

Cylinder

$r_\theta = D/2$ and r_ϕ tends to infinity, hence $S_\theta = 2S_\phi = pD/2t$ i.e. the *thin cylinder equation*.

Sphere

$r_\theta = r_\phi = D/2$ and hence $S_\theta = S_\phi = pD/4t$

The sphere is an ideal end closure since the stresses are less than those in other shapes. In practice, however, the high degree of mechanical forming necessary makes it impractical except for very high-pressure vessels.

Ellipsoid

An elliptical meridian of semi-major and -minor axis a, b, and eccentricity $e = 1 - (b/a)^2$ (see Fig. 3.6) is rotated about the minor axis to form the head of the cylinder with diameter $D = 2a$. The location of an element on the ellipse is defined most directly by the radius r from the rotation axis. However, it is usually convenient to define the alternative independent variable $u = [1 - e\,(r/a)^2\,]^{1/2}$ where $u = f(r)$ and $b/a \le u \le 1$. The geometry of the ellipse is then used to derive the radii of interest in terms of u (i.e. in terms of r).

$$r_\phi = (a^2/b)\, u^3 \;;\; r_\theta = (a^2/b)\, u$$

The stresses at the r element follow immediately from previous equations giving

$$S_\phi = (pa^2/2bt)\, u;\quad S_\theta = S_\phi\,(2 - 1/u^2) \text{ where } u = f(r)$$

Figure 3.6 shows the graphical result for $a = 2b$, the most common proportions for practical ellipsoidal vessel ends. Note how the prominent feature of this stress pattern is the tensile-to-compressive transition of the hoop stress at a diameter of about 0.8 D.

The consequences of the resultant compressive behaviour in the vessel are:
- a tendency towards local buckling;
- an increase in the equivalent stress (assuming the maximum shear stress theory, for example) with the third principal (radial) stress being zero (see Chapter 10 which covers theories of failure);
- incompatibility between the cylinder and its head, due to different senses of the hoop (i.e. diametral) strain. Under internal pressure the cylinder tends to expand, while the ellipsoidal end tries to contract diametrically at the head/shell junction. Bending moments in the walls of the cylinder and

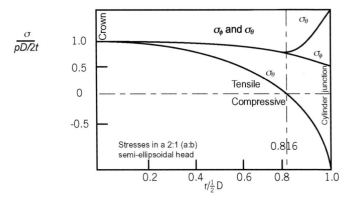

Tensile – compressive transition of hoop stress

A torispherical head

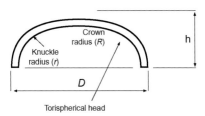

Fig. 3.6 Stresses in a semi-ellipsoidal vessel head

head are, therefore, set up at the junction, with corresponding bending stresses. These relatively minor secondary stresses cannot be explained by membrane theory, but empirical results show that they become insignificant at a distance of about five times the wall thickness from the junction. Most vessel codes state that a vessel should have a smooth transition (approx 4:1 taper) where the shell is welded to the head.

Torisphere

A torisphere is another common shape for a dished end (see Fig. 3.6). This consists of a spherical central portion of 'crown' radius, R, and a toroidal knuckle of radius r, where R/r is approximately twelve and R is about 95 per cent of the cylinder diameter. Junctions of the torus with both sphere and cylinder give rise to geometric singularities and hence to secondary bending stresses. For the torisphere, $S \geq M pR/2t\eta$

$M = [3 + (R/r)^{1/2}]/4$

where M = a stress concentration factor (specified in a vessel code).

Broadly, the greater the deviation from a sphere (where $R/r = 1$), the larger the M factor. The highly stressed region extends outwards from about 80 per cent D. Torispherical ends are often preferred to ellipsoidal ones, since the depth of drawing is less and so they are slightly cheaper to manufacture. This may be outweighed, however, by their higher stress concentrations and lower allowable pressure for a given material size.

3.2.6 Openings and compensation

The various types of openings that are needed by vessels tend to weaken the shell. Compensation, or *reinforcement*, is the provision of an extra stress-transmitting area in the wall of a cylinder or shell when material has been removed. Figure 3.7 shows the principles included. Figure 3.7 (i and ii) shows part of a cylinder's longitudinal section with the major circumferential stress acting across the critical longitudinal plane. The nominal material thickness is T and a hole of diameter D_b is bored to accommodate a branch connection. Dimensions are assumed to be relevant to the fully corroded condition. The stress-transmitting area removed is $A = D_b t$ where t is 'calculated' material thickness.

Figures 3.7 (iii) and 3.8 show how compensation is made for the area removed by providing an equal area for alternate force paths in otherwise unused material of the cylinder and branch. Not all the branch wall can be devoted to compensation since the internally pressurized branch is a cylinder in its own right, with calculation and nominal thicknesses, t_b and T_b, determined in a manner identical to the main shell. Provided that the longitudinal welds in both shell and branch do not lie in the critical

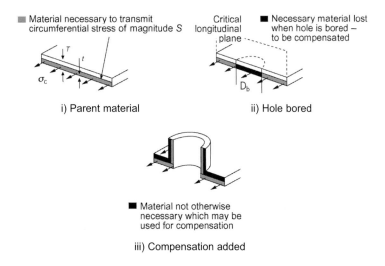

Fig. 3.7 The principles of compensation

longitudinal plane then, from a compensation point of view, both t and t_b would be calculated using a weld efficiency factor $\eta = 1$.

The thickness differences $T - t$ and $T_b - t_b$ contribute to compensation, but reinforcement tends to be less effective beyond the limits L_n normal to the vessel wall, and L_p from the branch centreline parallel to the wall. Figure 3.8 illustrates this for a set-in branch.

One national pressure vessel code gives the limits as

L_n = maximum [0.8 $(D_b T_b)^{1/2} + T_r$, minimum $(2.5T, 2.5T_b + T_r)$]
or $(D_b T)^{1/2}$ for a flanged-in head
L_p = maximum $(D_b, D_b/2 + T + T_b + 2c)$

In some design methods, the first of the L_p limits, i.e. D_b, is the controlling one. There is a general 'rule of thumb', however, that a compensating area cannot contribute to more than one branch. Therefore, if the spacing of two branches D_{b1} and D_{b2} is less than $D_{b1} + D_{b2}$, then by proportion $L_{p1} = D_{b1}$. spacing/$(D_{b1} + D_{b2})$. Also, if a branch is attached to a dished end, then no compensation area is effective if it lies outside an 80 per cent limit.

If the head is torispherical, the hole is generally located in the spherical portion and t will be given by the appropriate equation. If the head is ellipsoidal, some codes define an equivalent sphere for the application of the

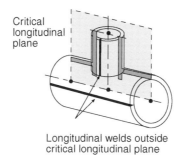

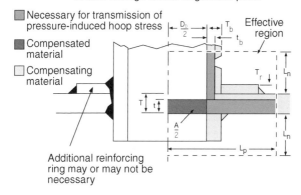

Fig. 3.8 Compensation – further details

appropriate equation, since the hole will not lie close to the rim region of sharp curvature, which dictates the thickness via a stress concentration factor.

Within the L_n, L_p limits, a general rule is that

$A_1 + A_2 + A_3 + A_4 + A_5 \geq A = D_b t$

As a general principle, compensation is normally positioned symmetrically about a hole and as close to the hole as possible. It is usually more economical to increase the branch thickness than to provide a separate reinforcing ring.

By necessity, the above principle of compensation is simplified and ignores inevitable stress concentrations, hence the need for substantial safety factors. Vessel codes often lay down branch size upper limits beyond which this simplistic approach is no longer permissible. Typically, one code specifies

Maximum branch bore = minimum ($D_i/2$, 500 mm) if $D_i \leq 1500$ mm,
or
$$= \text{minimum } (D_i/3, 1000 \text{ mm}) \text{ if } D_i > 1500 \text{ mm}$$

A general rule

As a 'rule of thumb' a single opening, 75 mm diameter or less in plate 6 mm thick or greater, does not require compensation and short pipes of this size are sometimes welded straight into a vessel shell.

3.2.7 Inspection openings

Various vessel openings are required in vessels to allow internal examinations as may be required by, for example, The Pressure System Safety Regulations (see Chapter 12). The size and disposition of the openings depend upon the duty and size of the vessel. In a small vessel a single handhole or a flanged-in inspection opening may be adequate, whereas large vessels require elliptical access manholes, often with reinforcement/seating rings. Alternatively, heads may be flanged inwards (reverse knuckle) to provide a seating surface (see Fig. 3.9).

The opening is sealed usually by an internal door, a gasket and one or two bridges (or dogs), and studs. The door is elliptical to permit its removal if necessary, for remachining of a damaged gasket seating surface. The minor axis of an elliptical opening in a cylindrical shell generally lies parallel to the longitudinal axis of the shell. The studs provide the initial sealing force, i.e. the initial seating pressure on the gasket face before the fluid is pressurized. When the fluid pressure later rises, the door tends to be self-sealing as the pressure load on the door increases the gasket contact pressure. The load on the studs then decreases. However, some vessel codes specify that the door must withstand simultaneous bending by both fluid pressure and maximum possible stud (or bolt) tightening.

A typical flat door calculation thickness, t, is determined from the generic form of

(C_1. fluid pressure. door area + C_2. bolt stress . bolt area)/$t^2 \leq S$;
where C_1, C_2 are constants.

Some vessel doors are equipped with a locating spigot to aid their engagement when closing. If the door is heavy then provision must be made for supporting it during opening or closing, any such support must not interfere with even take-up of the gasket, nor must it hinder easy access to the vessel.

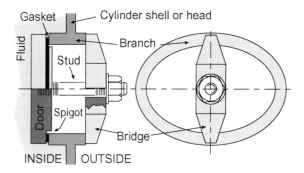

Fig. 3.9 A vessel inspection opening

3.2.8 Pipes and flanges

Long pipe runs that convey fluids between items of plant such as pressure vessels, are usually made up from short lengths of pipe welded together. The pipes are not always welded permanently into the vessels but may instead be attached in a manner that permits easy separation. Figure 3.10 shows general pipe flange types.

The most common, demountable connection involves welding a flange to each of the pipes, then connecting the flanges by nuts and bolts. A gasket material is interposed between the flanges to fill any imperfections in their faces, thereby preventing fluid leakage. The higher the fluid pressure and temperature, the more robust must be the flanges and bolting to contain them. Published standards contain tables of dimensions for flanges and other pipe fittings. The most common presentation is to list, for the various nominal pipe sizes, the minimum dimensions that are suitable for a certain limiting combination of fluid pressure and temperature. When steel flanges and fittings have to be selected for a design pressure and elevated temperature, their size is dictated by individual code tables (e.g. ANSI B16.5).

3.2.9 Pads

A small bore pipe is often attached to a vessel by means of a flange, butt-welded to the pipe and mating with a pad that is welded into, or formed on, the vessel wall [see Fig. 3.10 (iii)]. The pad dimensions must match the flange, and studs are screwed into tapped holes in the pad. Acceptable pad forms are defined in individual vessel codes.

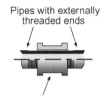

i) Internally threaded coupling

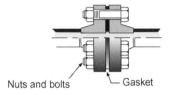

ii) Weld neck flanged

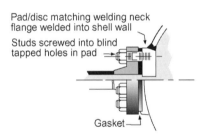

iii) Weld neck flange and pad

Fig. 3.10 General pipe-flange types

3.2.10 Vessel supports

Horizontal pressure vessels (length L, diameter D mm) are commonly mounted on two saddle supports. More than two supports would result in static indeterminacy and cause difficulty in predicting the load distribution in the event of foundation settlement. Supports generally extend at least 120 degrees around the vessel in order to transmit the reaction gradually into the shell wall. One support is attached to the vessel to prevent axial movement. The other is not attached but merely supports the vessel's weight, thus permitting free, longitudinal, thermal expansion.

The spacing of vessel supports has to take into account the actual bending mode of the vessel resulting from self-weight and fluid-weight. A 'rule of thumb' simply supported beam assumption suggests that the supports should be located at $0.207\,L$ from the ends, corresponding to bending moment magnitudes of $0.0214\,wL^2$ at both the centre and at the supports. However, this model neglects possible distortion that may occur when concentrated loads are applied to the relatively thin shell of a pressure vessel. To lessen this possibility, supports are more often located within $D/4$ of the ends to take advantage of the stiffening effect of the heads, although this does lead to bending stresses larger than those arising from the optimum simply supported beam assumption.

3.3 Simple pressure vessels (SPVs) – basic design

A simple horizontal or vertical pressure vessel, as the name implies, is very basic, comprising the following parts (see Fig. 3.11):
- cylindrical shell;
- two dished ends;
- a number of openings (inspection, drainage, inlet, outlet);
- nameplate and fixing.

Less 'simple' pressure vessels may have conical sections, flat ends, and saddle supports, and may be subjected to external as well as internal pressure. A typical technical standard is EN 286 (see also details of SPV legislation and standards in Chapter 11).

3.3.1 Material selection

SPVs are mainly manufactured of the following three types of material:
- ferritic steel (low carbon steel);
- austenitic steel (stainless steel);
- aluminium and its alloys.

The main ferritic steels used are:

EN 10207: Part 1	– Grades	SPH 235
		SPH 265
		SPHL 275
EN 10028	– Grades	PH 235
		PH 265
BS 1501	– 151 (360, 400, 430)	
	– 161 (360, 400, 430)	

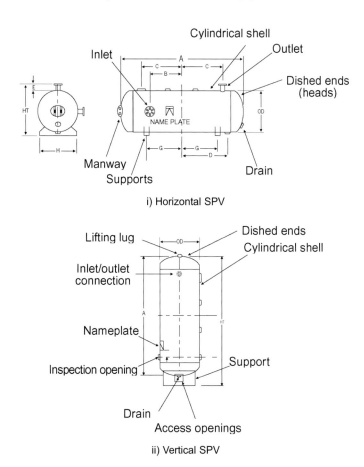

Fig. 3.11 Typical vertical and horizontal Simple Pressure Vessels (SPVs)

These are low carbon steels with restrictions on carbon, sulphur, and phosphorus content.

3.3.2 Welds

An SPV will typically contain the following welds (see Fig. 3.12):
- longitudinal welds (butt);
- circumferential welds (butt);
- nozzle or branch connection welds;
- fillet welds.

3.3.3 Stress calculations

SPV calculations are derived from thin-walled cylinder formulae. The following notation is commonly used:

P Internal pressure;
L Length of cylinder;
D (Mean) diameter of cylinder;
D_o (Outer) diameter of cylinder;
t 'Calculated' thickness of cylinder;
f Allowable design stress of the material (i.e. based on σ_θ);
σ_x Longitudinal stress;
σ_θ Circumferential stress (hoop stress).

The resulting stress formulae are:

$$\sigma_x = \frac{PD}{4t}$$

$$\sigma_\theta = \frac{PD}{2t}$$

These indicate that the hoop stress is twice the magnitude of the longitudinal stress and, as such, tends to be the 'controlling' stress in SPV design.

When determining the minimum thickness of a cylindrical shell the effects of corrosion have to be considered. The effects of corrosion are expressed as a corrosion allowance, which is typically 1 mm, i.e. for the life of a vessel it is anticipated that the shell wall thickness will reduce by 1 mm. The corrosion allowance is usually denoted by the letter c. Simple pressure vessel standard BS EN 286-1 specifies a 1 mm corrosion allowance. In practice additional factors are incorporated to allow for thickness tolerances,etc. There are three possible types of dished end or 'head' for SPVs: ellipsoidal, torispherical, or hemispherical. Calculations for determining the calculated thickness, t, of dished ends are based on thin-walled pressure vessel calculations for spherical shells.

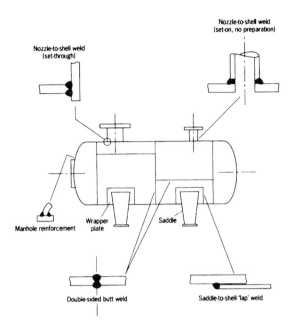

Fig. 3.12 Vessel weld types

For torispherical ends and ellipsoidal ends the method is to use *'design curves for dished ends'*, which are a graphical representation of the above formulae, modified slightly for ellipsoidal and torispherical ends. The steps are outlined below.

- First, the code limits for dished ends should be satisfied. These are:
  a) hemispherical ends : $\quad 0.002 D_o \leq t \leq 0.16 D_o$
  b) ellipsoidal ends : $\quad 0.002 D_o \leq t \leq 0.08 D_o$
  $\quad\quad\quad\quad\quad\quad\quad\quad\quad h_e \geq 0.18 D_o$
  c) torispherical ends : $\quad 0.002 D_o \leq t \leq 0.08 D_o$
  $\quad\quad\quad\quad\quad\quad\quad\quad r \geq 0.06 D_o$
  $\quad\quad\quad\quad\quad\quad\quad\quad r \geq 3t$
  $\quad\quad\quad\quad\quad\quad\quad\quad R \leq D_o$

  where R = crown radius and r = knuckle radius (see Fig. 3.6).

- Next the value h_e is determined (see the standard EN 286 for definitions). h_e is the smallest of the following:

$$\frac{h}{D_o^2} \over [4(R+t)]$$

$$\sqrt{\{D_o(r+t)/2\}}$$

- Once h_e is determined the following are calculated:

$$\frac{h_e}{D_o} \qquad \frac{PS}{10f}$$

The value t/D_o is the read off the design curves.
- As per the calculations for cylindrical shells the following equation is performed:
nominal plate thickness, e, $\geq t$ + corrosion allowance, c, + plate tolerance
(See the design curves for dished ends in BS EN 286-1: 1991 – Fig. 10.)

3.4 Gas cylinders – basic design

Gas cylinders, known generically as Transportable Pressure Receptacles (TPRs), are generally of forged or fabricated (two- or three-piece) construction. Figure 3.13 shows general arrangements. They are covered by their own directives and regulations and a wide range of published technical codes and standards.

Table 3.1 summarizes the major points of Gas Cylinder Design Standard EN 1442: 1998. Table 3.2 shows the manufacturer's marking requirements needed to comply with the standard. See also Tables 11.2 and 11.3 in Chapter 11 showing a summary of other major gas cylinder technical standards.

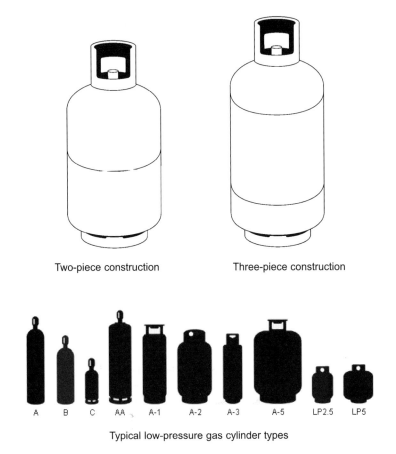

Fig 3.13 Gas cylinders – general design arrangements

Table 3.1 Gas cylinder standard EN 1442

Gas cylinder design standard BS EN 1442: 1998 Summary

Document title

Transportable refillable welded steel cylinders for Liquefied Petroleum Gas (LPG) – design and construction.

Materials (Section 4 of the Standard)
- Shell and dished-end materials have to comply with EN 10120.
- Materials for pressure-retaining parts need material certificates showing chemical analysis and mechanical properties.
- Material has to be 'identified' so that pressure-retaining parts can be traced to their origin.

Design (Section 5 of the Standard)

- Internal pressure, P_c, used for calculation purposes

  For butane cylinders; $P_c = P_{tmin} = 15$ bar
  For other LPG cylinders; $P_c = P_{tmin} = 30$ bar
  Where P_{tmin} = minimum permissible test pressure (bar)

- Cylindrical shell thickness, a

  $$a = \frac{P_c D}{\left(\frac{20 R_o J + P_c}{4/3}\right)}$$

  where D = cylinder OD (mm)
  R_o = minimum yield strength (N/mm^2)
  J = 0.9 for cylinders containing a longitudinal weld or 1.0 for cylinders without a longitudinal weld

  The result is subject to the following restriction on minimum shell wall thickness, a_{min}:
  for $D < 100$ mm $a_{min} = 1.1$ mm
  for 100 mm $\leq D \leq 150$ mm $a_{min} = 1.1 + 0.008 (D-100)$ mm
  for $D > 150$ mm $a_{min} = (D/250) + 0.7$ mm with an absolute minimum thickness of 1.5 mm

- Dished-end wall thickness (b_{min})

  $$b_{min} = \frac{P_c D C}{\left(\frac{20 R_o}{4/3} + P_c\right)}$$

  Where C is a shape factor that depends on the geometry of the dished end (see Table 1 and Figs 2 and 3 of the Standard).

Table 3.1 Cont.

Again, this is subject to minimum wall thickness which is the same as those used for shell thickness, *a*, above.
- Openings:
  - openings have to be located at one end only of the cylinder;
  - each opening needs a reinforcing pad or boss;
  - reinforcing pad welds must be clear of any circumferential shell welds.

Construction and workmanship

- Welding procedures must be approved to EN 288-3 and welders approved to EN 287-1.
- The longitudinal joint must be butt-welded; backing strips are not allowed.
- The circumferential joints (1 or 2) can be either butt-welded or 'joggled' (see Fig. 3.14).
- All welds need full penetration.
- Out-of-roundness tolerance $\left(OD_{max} - OD_{min}\right) = 0.01\left[\dfrac{OD_{max} - OD_{min}}{2}\right]$ for two piece cylinders or $0.015\left[\dfrac{OD_{max} - OD_{min}}{2}\right]$ for three-piece cylinders
- Straightness tolerance = maximum of 0.3 per cent of the cylindrical length.
- Cylinders must be heat treated (stress relieved) unless various specific criteria are all met (see Section 6.8.4 of the Standard).

Testing

- Mechanical test requirements vary depending on whether the cylinder is of two- or three-piece construction. Figure 3.15 shows the extent and location of tests required.
- Pressure tests are carried out to the pressure P_c as calculated above with membrane stress limited to 90 per cent of yield stress.
- A bursting test is required with the minimum bursting pressure, P_b, > 9/4 P_c and P_b > 50 bar.
- During the bursting test the following criteria have to be met:

$$\dfrac{\text{volumetric expansion}}{\text{initial cylinder volume}} \geq 20\% \text{ if cylinder length } L > \text{cylinder diameter } D$$

or $\geq 17\%$ if $L \leq D$

- When the cylinder bursts it must not fragment, show any evidence of brittle fracture, or reveal any material defects such as laminations.
- Radiographic testing of welds is required in accordance with EN 1435 class B (technique), ISO 2504 (assessment), and EN 25817 (defect acceptance criteria).

Table 3.1 Cont.

- Radiographic testing is required on the circumferential and longitudinal welds of the first production cylinder in a batch and one from each batch of 250 (see Fig. 3.16).
- The following indications (see EN 25817) are not permitted:
  - cracks, lack of penetration, or lack of fusion;
  - elongated inclusions;
  - groups of rounded inclusions of total length > 6 mm over a length of 12 × material thickness, a;
  - gas pores > (a/3) mm;
  - gas pores > (a/4) mm located ≤ 25 mm from any other gas pore;
  - gas pores where their total area > ($2a$) mm² over a length of 100 mm.
- Macro examination to EN 1321 of a transverse section of the cylinder welds.

Other key points

- The standard makes provision for cylinders to be manufactured under 'type-approval' (see EN 1442 Section 9).
- Cylinders have to be marked (see Table 3.2) by their manufacturer.

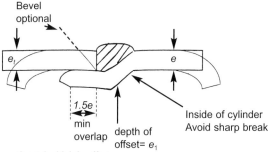

e) thickness of metal which is offset
e_1) thickness of metal which is not offset

Fig. 3.14 A gas cylinder 'joggled' butt joint

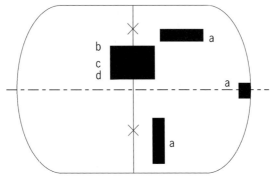

a) Alternative locations of test specimen for tensile test
b) Test specimen for tensile test
c) Test speciment for bend test (top side of the weld)
d) Test specimen for bend test (underside of the weld)

Test specimens taken from two-piece cylinders

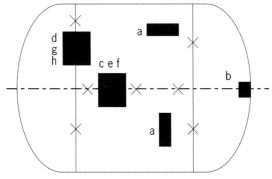

a) Alternative locations of test specimen for tensile test
b) Test specimen for tensile test
c) Test specimen for tensile test
d) Test specimen for tensile test

e) Test speciment for bend test (top side of the weld)
f) Test specimen for bend test (underside of the weld)
g) Test specimen for bend test (underside of the weld)
h) Test specimen for bend test (underside of the weld)

Test specimens taken from three-piece cylinders

Fig. 3.15 BS EN 1442: gas cylinders – mechanical test requirements

Extent of radiography of welds: cylinders with circumferential welds only

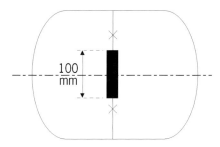

Extent of radiography of welds: cylinders with circumferential and longitudinal welds

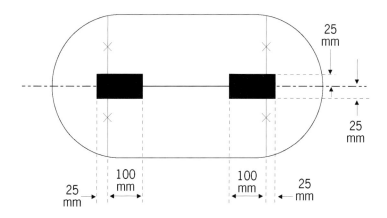

Fig. 3.16 BS EN 1442 gas cylinders – radiographic requirements

Table 3.2 Manufacturer's marking for gas cylinders (EN 1442: 1998). (Source: EN 1442: 1998 Table A.1)

	Definition	Example
1	European standard number.	EN 1442
2	For a cylinder that is normalized. This symbol is stamped immediately after the European standard number.	N
3	For a cylinder that is stress relieved. This symbol is stamped immediately after the European standard number.	S
4	Country of origin/manufacturer.	AXY
5	Manufacturing serial number; number to clearly identify the cylinder.	765432
6	Test pressure: the prefix P_t followed by the value of the test pressure. Measured in bar.	P_t 30 bar
7	Inspection stamp of competent body.	–
8	Test date: year (last two figures) and month of testing.	99/10
9	Water capacity: the minimum water capacity in litres guaranteed by the cylinder manufacturer followed by the unit '1'.	26, 21
10	Tare weight – marked on cylinder valves or on permanently attached fittings, where filling is by weight. The tare weight is the sum of the empty weight, the mass of the valve including a dip tube where fitted, and the mass of all other parts that are permanently attached to the cylinder when it is being filled (e.g. fixed valve guard). As an alternative, the requirement for the indication of tare weight is considered to be satisfied where the gross mass of the filled cylinder, the nature of the contents, and filling mass are marked.	–
11	Enough space for requalification date to be provided.	–
12	Where the cylinder is designed for commercial butane.	'Butane'
13	Space for additional stamp markings as required by the customer.	–

3.5 Heat exchangers – basic design

The basic definition of a heat exchanger is *a device that transfers heat from a hot to a cold fluid*. They are used extensively in process plants and are given specific names when they serve a special purpose, e.g. superheaters, evaporators, condensers, deaerators, etc., may all be classified as heat exchangers. For design purposes, heat exchangers can be broadly classified into *direct contact* and *surface types*. Figures 3.17 and 3.18 show the principles and the general extent of temperature changes associated with each type.

3.5.1 Contact-type exchangers

Contact-type exchangers are used mainly in steam systems where either steam is used as a contact heating medium or water is used as a cooling (i.e. attemporation or desuperheating) medium. The entire exchanger vessel is normally built to pressure equipment standards.

3.5.2 Surface-type exchangers

In surface-type exchangers, the two process fluids are kept separate by a physical barrier. Heat is transferred from the warm fluid through the barrier to the cold fluid. The two basic arrangements are the tube type and the plate type.

The design of tube-type heat exchangers is covered extensively by the TEMA (Tubular Exchanger Manufacturers' Association, USA) technical standards and European standards such as EN 247. Many configurations are available, divided broadly into *head* and *shell* types (see Fig. 3.19). Thermal expansion is accommodated by various types of sliding arrangements (Fig. 3.20). Tube-type exchangers show various different design arrangements for the way that the tubes are distributed in the tubeplate. Figure 3.21 shows the three most common – the triangular, square, and rotated-square patterns. Figure 3.22 shows an exploded view of a typical tube-type exchanger.

3.5.3 Thermal design

From a thermal viewpoint, tube-type heat exchangers can be classified broadly into parallel and counterflow types. The thermal 'driving force' is the parameter known as Log Mean Temperature Difference (LMTD). For the parallel flow configuration (see Fig. 3.17).

$$\text{LMTD}\left(\theta_m\right) = \frac{\theta_1 - \theta_2}{\ln\left(\theta_1 / \theta_2\right)} \text{ and } q = UA\theta \text{ (Watts)}$$

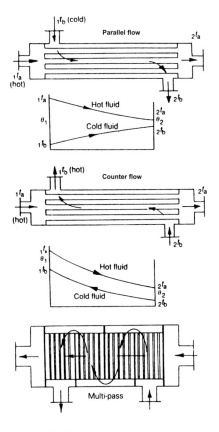

Fig. 3.17 Heat exchangers – principles

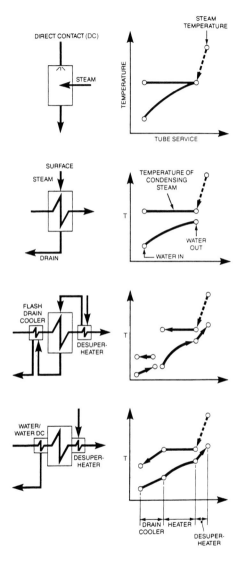

Fig. 3.18 Heat exchangers – principles (cont.)

Heat exchange in various types of heater. The diagrams illustrate the general temperature changes associated with various types of heater.

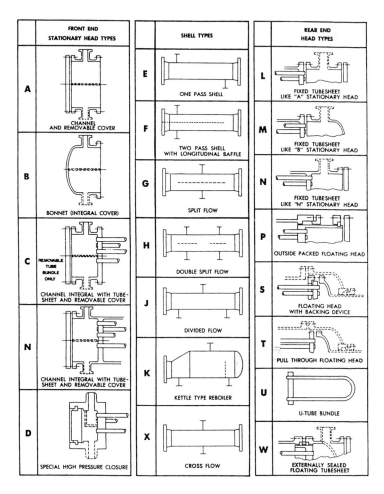

**Fig. 3.19 Surface-type exchanger configurations.
(Source: TEMA)**

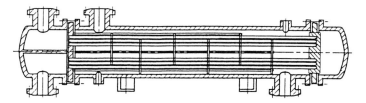

Sliding head and O-ring seals

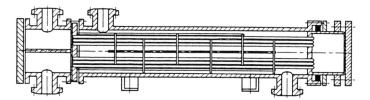

Sliding head and caulked seals

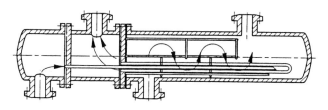

Bayonet type

Fig. 3.20 Surface-type heat exchangers – various arrangements. (Source: EN247: 1997)

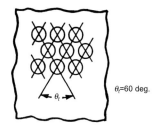

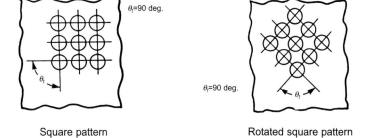

Fig. 3.21 Surface-type heat exchangers – tube patterns. (Source: TEMA)

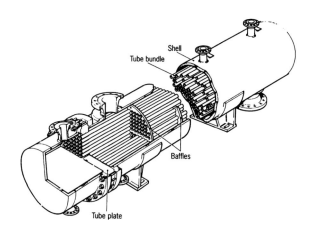

Fig. 3.22 Tube-type exchanger – exploded view

where
- A = surface area (m²)
- θ = temperature difference (°C)
- U = overall heat transfer coefficient (W/m²K)

The same formulae are used for counterflow exchangers, while for more complex crossflow, and multi-pass exchangers, LMTD is normally found from empirically derived tables.

3.5.4 Special applications

There are many variants of heat exchanger design used for specialized applications in power and process plants. Figures 3.23–3.25 show some specific examples.

3.6 Design by Analysis (DBA) – pr EN 13445

Design by Analysis (DBA) is an alternative philosophy to the historical method of Design by Formula (DBF). The idea behind DBF is that pressure equipment items (mainly vessels) are designed by reference to a set of formulae and charts included in codes and standards – this applies to the majority of vessels that are manufactured. The principle of DBA is that the design of a vessel can be checked or proven *individually* by a detailed technical investigation of the way the structure is loaded. A DBA methodology has been included in recent editions of the ASME pressure vessel and boiler code and is now being incorporated into the new proposed European standard on unfired pressure vessels – pr EN 13445.

3.6.1 What does DBA offer?

The basic reason for including the DBA methodology in pr EN 13445 is as a complement to the DBF approach where:
- superposition of loads have to be considered (wind, snow, and seismic loadings);
- there are fitness-for-purpose issues relating to cases when manufacturing tolerances are exceeded;
- very detailed analysis and 'design justification' is required for safety cases, etc.

In addition, it can act as a straight technical alternative to the DBF approach – typically for unusually shaped vessels that do not easily fit into standard vessel classes and categories.

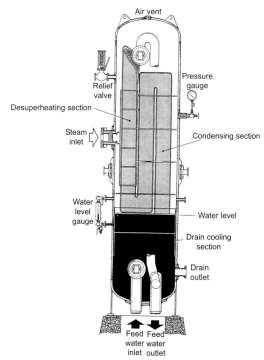

Fig. 3.23 A high-pressure feed heater with integral desuperheating and drain cooling sections

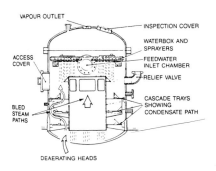

Fig. 3.24 A vertical 'contact-type' deaerator

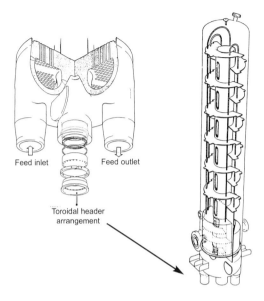

Fig. 3.25 A high-pressure feed heater with toroidal header

3.6.2 How does DBA fit into pr EN 13445?

The new proposed European standard on unfired pressure vessels will probably include two entries related directly to DBA:
- pr EN 13445-3, Annex B; *Direct route for Design by Analysis*
- pr EN 13345-3, Annex C; *Stress categorization route for Design By Analysis*

The emphasis is likely to be placed on providing general rules to allow checks to be made on the admissibility of a design, or confirmation of the maximum pressure (or other) load.

3.6.3 DBA – the technical basis

The basis of DBA is to guard against the following eight pressure vessel failure modes:
- elastic deformation (including elastic instability);
- plastic deformation;
- brittle fracture;
- stress rupture/creep deformation (inelastic);

- plastic instability/incremental collapse;
- high strain/low cycle fatigue;
- stress corrosion (cracking);
- corrosion fatigue.

A key point is that some of these failure mechanisms involve *plastic*, as well as elastic, deformations.

Stress categorization

An important part of the DBA methodology is the *categorization* (also termed *classification*) of stresses in a pressure vessel, so that they have proper relevance to the various failure modes that are considered. Table 3.3 shows the basic categories. (Note the similarity with the earlier ASME VIII (Div 2) approach mentioned in Chapter 4.)

Table 3.3 DBA stress categories

- Peak stress (F)
- Primary stress (P)
  - General primary membrane stress (P_m)
  - Local primary membrane stress (P_L)
  - Primary bending stress (P_b)
- Secondary stress (Q)

Stress limits

The principle of DBA is that designs incorporate *primary stress limits* to provide a suitable factor of safety against failure. Allowable stresses are expressed in terms of design stress, S_m, which is based on a consideration of the yield strength, R_e, and ultimate tensile strength, R_m, of the material. Normally S_m is taken as the design yield strength $\sigma_Y \times 2/3$. Table 3.4 outlines pr EN 13445 allowable stresses for primary and secondary stress combination in terms of S_m and σ_Y.

Table 3.4 pr EN 13445 *allowable stresses

	Allowable stress	*Stress intensity*
General primary membrane, P_m	k S_m	2/3 k σ_Y
Local primary membrane, P_L	1.5 k S_m	k σ_Y
Primary membrane plus bending $(P_m + P_B)$ or $(P_L + P_b)$	1.5 k S_m	k σ_Y
Primary plus secondary $(P_m + P_B + Q)$ or $(P_L + P_b + Q)$	3 S_m	2 σ_Y

* Provisional data

CHAPTER 4

Applications of Pressure Vessel Codes

4.1 Principles

Most of the international pressure vessel codes have been developed to the point where there is a high degree of technical similarity between them. Core areas such as vessel classes, design criteria and requirements for independent inspection, and certification are based on similar (but not identical) guiding principles. Vessel codes are increasingly being adopted for use on other types of pressure equipment, and their parameters for allowable stresses and factors of safety are used for design guidance. This is often the case for pressure equipment items that have thick cast sections, such as steam turbine casings and large cast valve chests, or those having similar construction to pressure vessels, such as high-pressure tubed heaters and condensers. In applications where pressure loading is not such a major issue the inspection and testing parts of vessel codes are sometimes still used. The welding and NDT requirements of both PD 5500 and ASME VIII/ASME V, for example, are frequently specified for gas dampers and ductwork, structural steelwork, tanks, and similar fabricated equipment. Changes are occurring as a result of the implementation of European directives and harmonized technical standards for various types of pressure equipment (see Fig. 11.1 in Chapter 11).

4.2 Code compliance and intent

The basic premise behind a pressure vessel code is that equipment should meet its requirements in all respects, – i.e. full *code compliance*. This is not always strictly adhered to, however, with the result that some equipment may be specified as being *'code intent'*. This means that it may comply with the code in some areas, e.g. design stresses, but not in others, such as the requirements for material traceability, NDT, and defect acceptance criteria. Most vessel and other pressure equipment codes are often seen to be applied in this way. Note that equipment built to such code intent cannot be officially 'code stamped'. Similarly, with the developing European harmonized standards, equipment can only carry the presumption of conformity with the underlying European directives [and hence compliance with the essential safety requirements (ESRs)] if it complies fully with requirements of the standard – partial compliance will have no real significance (see Chapter 11).

4.3 Inspection and test plans (ITPs)

The use of ITPs is well accepted throughout the pressure equipment manufacturing industry. The ITP is one of the most useful *working* documents. Used in a key monitoring and control role, it summarizes the activities of manufacturer, contractor, and certifying or notified body (see Chapter 11). Table 4.1 shows typical *technical* steps included in an ITP.

The following guidelines should be noted when dealing with pressure vessel ITPs.

- **Code compliance** A good pressure vessel ITP will refer to the relevant sections of the applicable code for major topics such as welding procedures, production test plates, and NDT.
- **Acceptance criteria** A reference to the defect acceptance criteria to be used should be shown explicitly in the ITP.
- **Hold points** The use of 'hold points', where a manufacturer must stop manufacture until an inspector or Notified Body completes an interim works inspection, should be shown.
- **Documentation review** Manufacturers compile the documentation package for a vessel after that vessel has successfully passed its visual/dimensional inspection and hydrostatic test. Typically, the package is compiled over a period of some two or three weeks after the hydrostatic test, during which time the vessel is shot-blasted, painted, and packed ready for shipping.

Applications of Pressure Vessel Codes 61

Table 4.1 Pressure Vessel ITP – typical content

	Activities	*Relevant documentation*
1	**Design appraisal (if applicable)**	Certificate from independent organization that the vessel design complies with the relevant code
2	**Material inspection at works (forgings, casting, plate, and tubes)**	Identification record/mill certificate (includes chemical analysis)
2.1	Identification/traceability	Witness identification stamps
2.2	Visual/dimensional inspection	Test report (and sketch)
2.3	UT testing of plate	Test certificate
2.4	Mechanical tests	Mechanical test results (including impact)
3	**Marking off and transfer of marks**	Material cutting record (usually sketches of shell/head plate and forged components)
4	**Examination of cut edges**	MPI/DP record for weld-prepared plate edges
5	**Welding procedures**	
5.1	Approve weld procedures	WPS/PQR records
5.2	Check welder approvals	Welder qualification certificates
5.3	Verify consumables	Consumables certificate of conformity
5.4	Production test plates	Location sketch of test plate
6	**Welding**	
6.1	Check weld preparations	Check against the WPS (weld procedure specification)
6.2	Check tack welds and alignment of seams/nozzle fit-ups	Record sheet
6.3	Back chip of first side root weld, DP test for cracks	DP results sheet
6.4	Visual inspection of seam welds	WPS and visual inspection sheet
7	**Non-destructive testing before heat treatment**	
7.1	RT or UT of longitudinal and circumferential seams	RT or UT test procedure

Table 4.1 Cont.

7.2	RT or UT of nozzle welds	Defect results sheet
7.3	DP or MPI of seam welds	NDT location sketches
7.4	DP or MPI of nozzle welds	
7.5	Defect excavation and repair	Repair record (and location sketch)
8	**Heat treatment/stress relieve**	
8.1	Visual/dimensional check before HT	Visual/dimensional check sheet
8.2	Heat treatment check (inc test plates where applicable)	HT time/temperature charts
9	**Non-destructive testing after HT**	
9.1	RT or UT of longitudinal and circumferential seams	RT or UT test procedures
9.2	RT or UT of nozzle welds	NDT results sheets
9.3	DP or MPI of seam welds	NDT location sheets
9.4	DP or MPI of attachment welds, lifting lugs, and jig fixture locations	
10	**Final inspection**	
10.1	Hydrostatic test	Test certificate
10.2	Visual and dimensional examination	Record sheet
10.3	Check of internals	Record sheet
10.4	Shot-blasting/surface preparation	Record sheet
10.5	Painting	Record of paint thickness and adhesion test results
10.6	Internal preservation	Record of oil type used
10.7	Vessel markings	Copy of vessel nameplate
10.8	Packing	Packing list
10.9	Documentation package	Full package, including index
10.10	Vessel certification	PD 5500 'Form X' or equivalent
10.11	Concession details	Record of all concessions granted (with technical justification)

4.4 Important code content

1. Responsibilities
Important points are:
- relative responsibilities of the manufacturer and the purchaser;
- the recognized role of the independent inspection organization;
- the technical requirements and options that can be agreed between manufacturer and purchaser;
- the way in which the certificate of *code compliance* is issued – and who takes responsibility for it.

2. Vessel design
Important points are:
- details of the different construction *categories* addressed by the vessel code;
- knowledge of the different classifications of welded joints;
- knowledge of prohibited design features (mainly welded joints).

3. Materials of construction
Codes contain information about materials for pressurized components of the vessel such as:
- materials that are *referenced* directly by the code – these are subdivided into plate, forged parts, bar sections, and tubes;
- the code's requirement for other (non-specified) materials, in order that they may be suitable for use as *permissible* materials;
- any generic code requirements on material properties such as carbon content, UTS, or impact value;
- specific requirements for low temperature applications.

4. Manufacture, inspection, and testing
The relevant areas (in approximate order of use) are:
- requirements for material identification and traceability;
- NDT of parent material;
- assembly tolerances (misalignment and circularity);
- general requirements for welded joints;
- welder approvals;
- production test plates;
- the extent of NDT on welded joints;
- acceptable NDT techniques;
- defect acceptance criteria;
- pressure testing;
- content of the vessel's documentation package.

4.5 PD 5500

PD 5500: 2000 'Specification for unfired fusion welded pressure vessels' is a recent replacement for British Standard BS 5500: 1997. It is issued under the status of a PD (published document) in anticipation of the forthcoming harmonized standard on pressure equipment that will be adopted for use in the UK. Table 4.2 gives an outline summary of PD 5500 content. Table 4.5 indicates how the technical content claims to satisfy the essential safety requirements (ESRs) of the Pressure Equipment Directive (PED) (see Chapter 11).

Tables 4.3, 4.4 and Figs 4.1 and 4.2 show key technical points.

Table 4.2 PD 5500 – a summary

Information you need	*How to find it in PD 5500*
1 Responsibilities	Section 1.4 places the responsibility for code compliance firmly on the manufacturer.
• Certification	Look at Section 1.1 – it confirms that PD 5500 applies only to vessels built under independent survey. Section 1.4 specifies that code compliance is documented by the use of 'Form X' – this is issued by the manufacturer and countersigned by The Inspecting Authority (the 'third party').
• Manufacturer/purchaser agreement	Table 1.5 lists approximately 50 points for separate agreement.
2 Vessel design	Vessel design is covered by Section 3 (the longest section) of PD 5500.
• Construction categories	Section 3.4 defines the three construction categories (1, 2, and 3) – they have different material and NDT requirements.
• Joint types	Section 5.6.4 defines types A and B welded joints, which have different NDT require-ments.
• Design features	Figures E.1 (1) to E.1 (6) show typical (rather than mandatory) weld joint details.
3 Materials	Materials selection is covered by Section 2.
• Specified materials	Section 2.1.2 (a) references the British Standard materials that are specified. The main ones are BS 1501* (plates), BS 1503* (forgings) and BS 3601* to 3606* (tubes).

Applications of Pressure Vessel Codes 65

Table 4.2 Cont.

•	Other permissible materials	\multicolumn{2}{l}{Section 2.1.2.1 (b) explains that other materials *are allowed* as long as they meet the criteria specified in this section, and in Section 2.3.2.}
•	Material properties	\multicolumn{2}{l}{Tables 2.3 give the specified material properties for all pressurized materials.}
4	**Manufacture, inspection, and testing**	*See PD5500 section*	*Summary*
•	Material identification	4.1.2	Material for pressure parts must be 'positively identified' [but note there is no reference to material certificate 'levels' (as in EN 10204)].
•	NDT of parent material	5.6.2	This is not mandatory (see 4.2.1.2 for NDT of cut edges).
•	Assembly tolerances	4.2.3	Alignment and circularity (out-of-roundness) are closely specified. Tolerances depend on the material thickness.
•	General requirements: welded joints	4.3	The manufacturer must provide evidence of 'suitable' welding.
•	Welder approvals	5.2/5.3	WPS and PQRs must be approved. Welders must be qualified.
•	Production test plates	5.4	Not mandatory for carbon steel vessels.
•	Extent of weld NDT	5.6.4	Purchaser option. Different extent for Cat 1, 2, 3 vessels, and type A or B welds.
•	NDT techniques	5.6.4	UT and RT testing are both acceptable volumetric NDT techniques.
•	Defect acceptance criteria	5.7	Use Tables 5.7 (1)–5.7 (4) depending on the technique used.
•	Pressure testing	5.8	A witnessed test is required. Section 5.8.5.1 gives the formula for calculating test pressure.

Table 4.2 Cont.

• Documentation package	1.4.4	Form 'X' is mandatory. Section 1.5.2.2 lists the minimum documentation required for code compliance.	

* Superseded by EN 100 xx series such as:

EN 10028: Plates
EN 10213: Castings
EN 10216: Seamless tubes
EN 10217: Welded tubes
EN 10222: Forgings

Table 4.3 PD 5500 Cat 1: Key points

1 Code compliance
- It is unusual for boiler drums not to be specified as fully compliant with Cat 1 requirements. Remember that independent survey and certification is mandatory for full PD 5500 compliance.

2 Materials of construction
- If 'non-specified' materials are used, do a full check on the material properties. PD 5500 Section 2 specifies the checks that are required for materials to be permissible.
- Check that all pressure-retaining material is positively identified and is traceable to its source. As a guide, high-alloy steels should have EN 10 204 3.1A certificates and other pressure parts should have 3.1B traceability – but these EN certificates are not mandatory.

3 Manufacturing inspection
- Check the weld preparations (after tack-welding) for the correct root gap and accuracy of alignment. Look carefully for distortion.
- Tolerances and alignments are important. Check the following:

Circumference:

for O.D < 650 mm, tolerance is + 5 mm

for O.D > 650 mm, tolerance is + 0.25 per cent of circumference

Straightness:

max deviation is 0.3 per cent of total cylindrical length

Circularity:

circularity is represented by ID max - ID min at any one cross-section. The maximum allowable is:

ID max - ID min < [0.5 + 625/O.D] to a maximum of 1 per cent

Applications of Pressure Vessel Codes

Table 4.3 Cont.

Surface alignment:

for material thickness (*e*)

Longitudinal joints, $e < 12$ mm, max misalignment is $e/4$

$12 \text{ mm} < e < 50$ mm, max misalignment is 3 mm

Circumferential joints, $e < 20$ mm, max misalignment is $e/4$

$20 \text{ mm} < e < 40$ mm max misalignment is 5 mm

4 Documentation

Final certification should use the 'Form X' shown in PD 5500 Section 1. Check that the correct wording has been used.

Table 4.4 PD 5500 Cat 2: Key points

1 Code compliance
- As with Cat 1 vessels it is unusual for Cat 2 pressure vessels not to be specified as fully compliant with code requirements. Independent survey and certification is mandatory.

2 Materials of construction
- PD 5500 Cat 2 vessels have limits placed on maximum plate thickness – this limits the design pressure.
- Material traceability requirements are the same as for Cat 1 vessels.

3 Manufacturing inspection
- Note that permanent weld backing-strips are allowed for Cat 2 vessels. Check that they are of the correct 'compatible' material.
- Alignment tolerances are the same as for Cat 1 vessels.
- Because only a 'percentage' volumetric NDT is required it is doubly important to make sure that critical areas are investigated. These are:
  - intersections of longitudinal and circumferential joints
  - areas of weld seam that are within 12 mm of nozzle openings
  - the ends of weld seams, particularly the 'beginning' of the run
- For radiographic examination make sure that the '10 per cent of aggregate length' requirement includes at least one radiograph from each weld seam. This is good practice.
- Hydrostatic test pressure is the same as for Cat 1 vessels.

4 Documentation
Final certification should use the 'Form X' shown in PD 5500 Section 1. Check that the correct wording has been used and that the 'Cat 2' designation is clearly indicated.

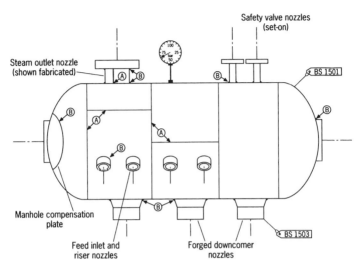

Materials
- Specified materials are given in PD 5500 section 2. Other materials can be permissible if specification and testing requirements are met
- No maximum plate thickness is specified
- There are temperature limitations for ferritic steels. Austenitic materials are permissible down to approx -200 °C

Welded joints
- Types A and B joints are as shown

	NDT requirements	
Location	NDT extent and technique	Acceptance criteria
Plate material	US examination to BS 5996 (optional)	BS 5996
Weld preparations	Visual examination is mandatory	By agreement
Type A welds	100% RT or UT is mandatory. Surface crack detection (DP or MPI) is optional	RT: PD 5500 Table 5.7 (1) UT: PD 5500 Table 5.7 (2)
Type B welds	100% RT or UT (some exceptions based on thickness) 100% DP or MPI, including attachment welds	As above PD 5500 Table 5.7 (3)
Test plate	Not mandatory	By agreement

Fig 4.1 Vessel to PD 5500 Cat 1

Applications of Pressure Vessel Codes 69

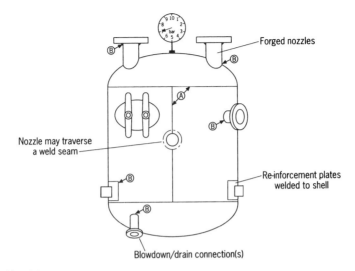

Materials
- Plate thickness is limited to: 40 mm for carbon steel (groups M0 and M1)
                                  30 mm for C–Mn steel (group M2)
                                  40 mm for austenitic steel
- Specified materials are the same as for Cat 1 vessels - see PD 5500 Tables 2.3

| | **NDT requirements** | |
Location	*NDT extent and technique*	*Acceptance criteria*
Plate material	UT examination to BS 5996 (optional)	BS 5996
Weld preparations	Visual examination is mandatory	By agreement
Seam welds (type A)	Minimum 10 % RT or UT of the aggregate length but including all intersections and seam areas near nozzles (as shown)	Use PD 5500 Tables 5.7 (1)–(4). There is a special re-assessment technique if defects are found, see PD 5500 Fig. 5.7
Nozzle welds (type B)	Full UT or RT on one nozzle per ten. 100% DP or MPI on all nozzle welds	PD 5500 Table 5.7 (1) or (2) Table 5.7 (3)
Compensation plate welds	100% DP or MPI	PD 5500 Table 5.7 (3)
Attachment welds to pressure parts	10% DP or MPI	PD 5500 Table 5.7 (3)
Test plate	Not mandatory	By agreement

Fig 4.2 Pressure vessel to PD 5500 Cat 2

4.5.1 PD 5500 and the PED ESRs

In recognition of the future requirement for a UK harmonized standard for pressure vessels, PD 5500 includes information showing how its technical content complies with the essential safety requirements (ESRs) of the pressure equipment directives (PED). Table 4.5 shows an outline of these comparisons. Note, however, that as PD 5500 does not currently have formal status as a harmonized standard, compliance with its technical requirements does not carry an *automatic* presumption of conformity with the PED.

Table 4.5 PD 5500 and the PED essential safety requirements

PED Annex 1 clause	PED essential safety requirement	PD 5500 reference
1	**General**	
1.1	Equipment to be designed, manufactured, checked, equipped, and installed to ensure safety, when put into service in accordance with manufacturer's instructions.	1
1.2	Principles to be used by the manufacturer in assessing appropriate solutions to:	Annex N.3
	– eliminate or reduce hazards;	
	– apply protection measures against hazards;	
	– inform user of residual hazard.	
1.3	Equipment to be designed to prevent danger from foreseeable misuse	1
	– warning to be given against misuse.	
2	**Design**	
2.1	Equipment to be designed to ensure safety throughout intended life and to incorporate appropriate safety coefficients.	3.1
2.2.1	Equipment to be designed for loadings appropriate to its intended use.	3.2.1
2.2.2	Equipment to be designed for adequate strength based on calculation method and to be supplemented, if necessary, by experimental methods.	3.2.2

Applications of Pressure Vessel Codes 71

Table 4.5 Cont.

2.2.3(a)	Allowable stresses to be limited with regard to foreseeable failure modes; safety factors to be applied. Requirements to be met by applying one of the following methods: – design by formula; – design by analysis; – design by fracture mechanics.	3.2
2.2.3(b)	Design calculations to establish the resistance of equipment, in particular: – calculation of pressure to take account of maximum allowable pressure, static head, dynamic fluid forces, and decomposition of unstable fluids; – calculation of temperature to allow appropriate safety margin; – maximum stresses and peak stresses to be within safe limits;	3.2
	– calculations to utilize values appropriate to material properties and to be based on documented data and have appropriate safety factors;	Annex K 3.4.2
	– use of joint factors; – to take account of foreseeable degradation mechanisms: instructions to draw attention to features that are relevant to the life of equipment.	
2.2.3(c)	Calculations to allow for adequate structural stability during transport and handling.	3.2.1
2.2.4	Experimental design method. Design to be validated by appropriate test programme on representative sample of equipment. Test programme to include pressure strength test to check equipment does not exhibit leaks or deformation, exceeding determined threshold.	3.2.2 and 5.8.6
2.3	Provisions to ensure safe handling and operation. Methods to be specified for operation of equipment to preclude any foreseeable risk with attention being paid to: – closures and openings; – discharge of pressure relief blow-off; – access while pressure or vacuum exists;	1

Table 4.5 Cont.

	– surface temperature;	
	– decomposition of unstable fluids;	
	– access doors equipped with devices to prevent hazard from pressure.	
2.4	Means of examination:	3.7.2.1, 3.12, 3.10.1.1
	– to be designed and constructed so that examinations can be carried out;	
	– able to determine internal condition;	
	– alternative means of ensuring safe conditions.	
2.5	Means to be provided for draining and venting where necessary:	1
	– to avoid water hammer, vacuum collapse, corrosion, and chemical reactions;	
	– to permit cleaning, inspection, and maintenance.	
2.6	Adequate allowance or protection to be provided against corrosion.	3.3
2.7	Adequate measures to be taken against effects of erosion or abrasion by design, replacement parts, and instructions.	3.3.4
2.8	Assemblies to be designed so that:	2
	– components to be assembled are suitable and reliable for their duty;	
	– components are properly integrated and assembled	
2.9	Assemblies to be provided with accessories, or provision made for their fitting, to ensure safe filling or discharge with respect to hazards:	1
	– on filling, by overfilling or over-pressurization and instability;	
	– on discharge, by uncontrolled release of fluid:	
	– on unsafe connection or disconnection.	
2.10	Protection against exceeding the allowable limits. Equipment to be fitted with, or provision made for, suitable protective devices, that are determined on the basis of characteristics of equipment.	5.13

Table 4.5 Cont.

2.11.1	Safety accessories to be designed and constructed to be reliable and suitable for intended duty, including maintenance; to be independent or unaffected by other functions.	3.13.1
2.11.2	Pressure limiting devices to be designed so that design pressure will not be exceeded except for short duration pressure surges of 1.1 times design pressure.	3.13.2
2.11.3	Temperature monitoring devices to have adequate response time on safety grounds.	2
2.12	External fire. Pressure equipment to be designed and fitted with accessories to meet damage limitation requirements.	1.1.5
3	**Manufacturing**	
3.1.1	Preparation of component parts not to give rise to defects, cracks, or changes in mechanical characteristics likely to affect safety.	4.2
3.1.2	Permanent joints and adjacent zones:	4.3
	– to be free from surface and internal defects detrimental to safety;	
	– properties to meet minimum specified for materials to be joined or taken into account in design calculations;	
	– permanent joining to be carried out by suitably qualified personnel according to suitable procedures;	
	– personnel and procedures to be approved by third party for Category II, II, and IV equipment.	
3.1.3	Non-destructive tests of permanent joints to be carried out by suitably qualified personnel.	5.6.1
	– Personnel to be approved by third party for Category II, III, and IV equipment.	
3.1.4	Suitable heat treatment to be applied at appropriate stage of manufacture.	4.4
3.1.5	Traceability. Material making up component parts to be identified by suitable means from receipt, through production, up to final test.	4.1.2

Table 4.5 Cont.

3.2.1	Final inspection to be carried out to assess, visually and by examination of documentation, compliance with the requirements of the Directive.	5.8
	– Tests during manufacture to be taken into account.	
	– Inspection as far as possible to be carried out internally and externally on every part	
3.2.2	Hydrostatic pressure tested to a pressure at least equal to value laid down in ESR 7.4.	5.8
	– Category I series equipment may be tested on a statistical basis.	
	– Other tests, of equivalent validity, may be carried out where hydrostatic is harmful or impractical.	
3.2.3	For assemblies, the safety devices to be checked to confirm compliance with ESR 2.10.	1
3.3	Marking. In addition to GE marking, all relevant information to be provided as listed in ESR 3.3.	1.4.4, 5.8.9
3.4	Operating instructions. All relevant instructions to be provided for the user as listed in ESR 3.4.	1
4	**Materials**	
4.1(a)	Materials to have appropriate properties for all operating and test conditions, to be sufficiently ductile and tough (refer to ESR 7.5).	2.2
4.1(b)	Materials to be chemically resistant to contained fluids.	3.3.1
	– Properties not to be significantly affected within scheduled lifetime of equipment.	
4.1(c)	Materials not to be significantly affected by ageing.	2
4.1(d)	Material to be suitable for intended processing (manufacturing) procedures	2
4.1(e)	Materials to avoid undesirable effects when joining.	4.2.2.1
4.2(a)	Manufacturer to define material values necessary for design and for requirements of ESR 4.1.	2
4.2(b)	Manufacturer to provide technical documentation relating to compliance with material specifications of PED in one of the following forms:	2
	– by using materials which comply with harmonized standards;	

Table 4.5 Cont.

	– by using materials covered by European approval in accordance with Article 11 of PED;	
	– by a particular material appraisal.	
4.2(c)	For equipment in PED categories III or IV, the particular material appraisal to be performed by notified body in charge of conformity assessment.	2
4.3	Manufacturer to take appropriate measures to ensure:	2
	– materials conform to specification;	
	– documentation by material manufacturer affirms compliance with a specification;	
	– documentation for main pressure parts in PED categories II, III, or IV to be a certificate of specific product control.	
5	**Fired or process heated pressure equipment**	1
5(a)	Means of protection to be provided to restrict operating parameters to avoid risk of local or general overheating.	1
5(b)	Sampling points to be provided for evaluation of properties of the fluid to avoid risks from deposits and/or corrosion.	1
5(c)	Adequate provisions to be made to eliminate risk of damage from deposits.	1
5(d)	Means to be provided for safe removal of residual heat after shutdown.	1
5(e)	Steps to be taken to avoid dangerous accumulation of combustible substances and air, or flame, blowback.	1
6	**Piping**	1
7	**Supplementary requirements to sections 1–6**	
7.1	Allowable stresses.	Annex
7.2	Joint coefficients	3.4.2
7.3	Pressure limiting devices. The momentary pressure surge to be kept to 10 per cent of design pressure.	Annex J

		Table 4.5 Cont.
7.4	Hydrostatic test pressure to be not less than the greater of:	5.8
	– maximum loading to which equipment may be subject in service, taking into account design pressure and design temperature, multiplied by 1.25; or	
	– design pressure multiplied by 1.43.	
7.5	Material characteristics. Unless other values are required in accordance with different criteria to be taken into account, a steel is considered sufficiently ductile to satisfy 4.1(a) if, in a tensile test carried out by a standard procedure, the following are met:	2
	– elongation after rupture is not less than 14 per cent;	
	– bending rupture energy measured on an ISO V test piece not less than 27 J at a temperature not greater than 20 °C but no higher than the lowest scheduled operating temperature.	
1	Outside the scope of the current edition of PD 5500.	
2	Requirements are not covered in the current edition of PD 5500.	

4.6 The ASME vessel codes

4.6.1 Summary

The ASME code comprises the following sections:
ASME Section
- I Rules for construction of power boilers
- II Material specifications
- III Rules for construction of nuclear power plant components
- IV Rules for construction of heating boilers
- V Non-destructive examination (NDE)
- VI Rules for care and operation of heating boilers
- VII Rules for care of power boilers
- VIII Rules for construction of pressure vessels
- IX Welding and brazing qualifications
- X Fibre-reinforced plastic pressure vessels
- XI Rules for in-service inspection of nuclear power plant components

ASME VIII

The ASME Boiler and Pressure Vessel Code, Section VIII is an accepted code used in the USA and many other parts of the world. It provides a thorough and basic reference for the design of pressure vessels and heat exchangers covering design, material selection, fabrication, inspection, and testing. Being a comprehensive code, ASME VIII is complicated and until you become familiar with its structure and principles, can be difficult to follow. Table 4.6 gives an outline summary. Figure 4.3 shows a quick reference guide to the major technical aspects of an ASME VIII vessel.

Table 4.6 ASME VIII – A summary

Information you need	How to find it in ASME Section VIII
1 Responsibilities	Part UG–90 lists 19 responsibilities of the manufacturer and 14 responsibilities of the Authorized Inspector (AI).
• Certification	For vessels to carry the ASME 'stamp' they must be constructed and inspected fully in compliance with the code. The manufacturer must have a 'Certificate of Authorization'.
	Parts UG–115–120 specify marking and certification requirements.
• Manufacturer/purchaser agreement	Although such agreements are inferred, there is no explicit list of 'agreement items'.
2 Vessel design	There are two 'divisions of vessel', Div 1 and Div 2. Most vessels will qualify under Div 1. Div 2 is for specialized vessels with restricted materials and onerous testing requirements.
• Construction categories	
• Joint efficiencies	UW–12
• Joint types	Part UW–3 defines category A, B, C, D welded joints, which have different NDT requirements.
• Design features	Figures UW–12, 13.1, 13.2, and 16.1 show typical acceptable (and some unacceptable) welded joints. Most of the information is for guidance, rather than being mandatory.
• Design pressure	UG–19, UG–21
• Max allowable working pressure	UG–98
• Design temperature	UG–19, UG–20

Table 4.6 Cont.

3 **Materials**	Part UG–4 references materials specified in ASME Section II.
• Permissible materials	
	Part UG–10 allows alternative, or incompletely specified, material to be recertified.
	Typical specified materials are: ASTM SA–202 (plates), SA–266 (forgings), SA–217 (castings) and SA–192 (tubes).

4 **Manufacture, inspection, and testing**	***See ASME VIII part***	***Summary***
• Material identification	UG–94	Material for pressure parts must be certificated and marked. This infers traceability.
• NDT of parent material	UG–93(d)	Visual examination is inferred. DP/MPI is only mandatory for certain material thickness and applications.
• Assembly tolerances	UG–80 and UW–33	Circularity (out-of-roundness) and alignment tolerances are closely specified.
• Welder approvals	UW–28, 29, 48	WPS, PQR, and welder qualifications are required.
• Production test plates	UG–84	Test plates are required under the conditions set out in UG–84.
• Extent of weld NDT	UW–11	UW–11 gives priority to RT techniques. UT is permitted in certain exceptional cases.
• NDT techniques	UW–51, 52, 53, UW–42, UW–11	ASME Section V techniques are specified.
• Defect acceptance criteria	UW–51, 52	UW–51 gives acceptance criteria for 100 per cent RT testing.
• Repairs	UG–78, UW–38, W–40(d), UF–37	UW–52 gives acceptance criteria for 'spot' RT testing.
• Pressure testing	UW–50, UG–99, UCI–99, UCL–52	A witnessed test is required. UG–99(b) specifies a test pressure of 150 per cent working (design) pressure.

Table 4.6 Cont.

• Documentation package	UG–115 to 120	Marking and documentation requirements are mandatory for full code compliance.
• Inspection	UG–90 to UG–97	
• Quality control systems	Appendix 10, UW–26	
• Proof tests	UG–101	

ASME VIII is divided into the Divisions 1 and 2. Division 1 (VIII–1) covers normal vessels. Division 2 (VIII–2) covers alternative rules for pressure vessels and is used for various types of special applications, including nuclear. Although VIII–2 was recently formally withdrawn, it is still used by many manufacturers. Both are structured into a large number of paragraphs designated by reference letters, e.g. UG–22.

ASME VIII is written against a well-defined, theoretical background which is similar, but not identical, to that used for other pressure vessel codes. This theoretical background is reflected in the design rules that apply to all the components of pressure vessels and the way in which size, shape, and material choices are made. Many design rules in VIII–1 and VIII–2 are similar and the important ones are summarized in the Appendices of the codes as:

- *Mandatory appendices* – These are design rules that have been well substantiated over time.
- *Non-mandatory appendices* – These are in a 'trial period' waiting to be transferred to the mandatory appendices once their safety and practicality in use have been validated.

The content of ASME Section VIII is organized to cover the following main design topics:

- allowable stresses and joint efficiencies;
- cylindrical shells under internal and external load;
- dished ends;
- flat plates, covers, and flanges;
- vessel openings;
- heat exchangers.

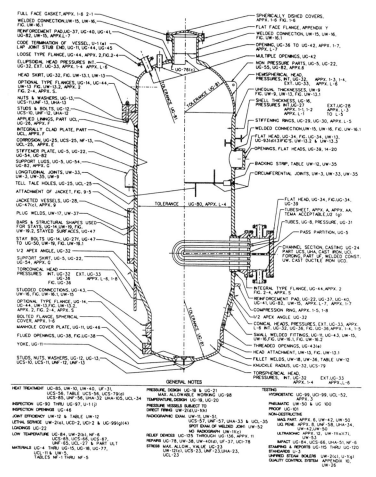

The purpose of this guide is to illustrate some of the types of presure vessel construction which are provided under section VIII, Division 1, of the ASME code to furnish direct reference to the applicable rule in the code. In the event of a discrepancy, the rules in the current edition of the code shall govern.

Fig. 4.3 ASME VIII Div 1 – quick reference guide

4.6.2 Allowable stresses

The ASME code works on the principle of *allowable stress*, i.e. the maximum stress that a vessel component will experience in use. Broadly speaking this is:
- maximum allowable stress = 1/4 × Ultimate Tensile Strength (UTS) or 2/3 × yield strength, whichever is the smaller. This applies only for temperatures below the creep range (nominally 650 °F/343.3 °C);
- for bolts, maximum allowable stress = 1/5 × UTS.

In all cases, the strength value used is the 'effective' strength; i.e. that at the design temperature of the component.

The '1.1 Factor'

ASME practice for 'effective strength' is to introduce a factor of 1.1 (based primarily on reference data and engineering judgment). Table 4.7 shows the details. For components operating in the creep range, ASME VIII broadly assumes that allowable stress is controlled by creep–rupture rather than straightforward tensile–yield criteria, and separate material strength tables and curves are used.

Joint efficiency factors (E)

ASME VIII–1 uses 'joint efficiency factors' (E). This is a stress multiplier applied to components of a vessel in which the welded joints have not been subject to 100 per cent Radiographic Testing (RT). Effectively, this means that VIII–1 vessels have variable factors of safety, depending on which parts are subject to RT and which are not. As a rule of thumb, longitudinally welded butt joints in cylindrical shells have:
- Fully radiographed: $E = 1$ = factor of safety of 4
- Non-radiographed: $E = 0.7$ = factor of safety of 5.7

Vessel welded joints are divided into four categories: A, B, C, D. Category A contains mainly longitudinal and head-to-shell circumferential seams. Category B encompasses mainly circumferential seams between shell sections, while flange-to-shell joints are generally category C. Joints in category D are the attachment of nozzles to the vessel shell, head, and any transition sections. Figure 4.4 shows these categories. Note that the categories refer to the locations of a joint in a vessel, rather than the nature of the weld construction.

4.6.3 Cylindrical vessel shells

ASME VIII–1 gives rules for the design of cylindrical shells of uniform thickness subject to internal pressure, external pressure, and axial loads. The governing equations for longitudinal and circumferential stresses are basically as shown in Table 4.8.

Table 4.7 ASME VIII-1 allowable stress values. (Source: ASME)

Product/material	Below room temperature			Room temperature and above			
	Tensile strength	Yield strength	Tensile strength	Yield strength	Stress rupture		Creep rate
Wrought or cast ferrous and nonferrous	$\dfrac{S_T}{4}$	$2/3\ S_y$	$\dfrac{1.1}{4} S_T R_T$	$2/3\ S_y$	$2/3\ S_y R_y$ or $0.9 S_y R_y$ [Note(1)]	$0.67 S_{Ravg}$	$0.8 S_{Rmin}$	$1.0 S_c$
Welded pipe or tube, ferrous and nonferrous	$\dfrac{0.85}{4} S_T$	$2/3 \times 0.85\ S_y$	$\dfrac{0.85}{4} S_T$	$\dfrac{(1.1 \times 0.85)}{4} S_T R_T$	$2/3 \times 0.85\ S_y$	$2/3 \times 0.85\ S_y R_y$ or $0.9 \times 0.85 S_y R_y$ [Note (1)]	$(0.67 \times 0.85)\ S_{Ravg}$	$(0.8 \times 0.85)\ S_{Rmin}$	$0.85 S_c$
Ferrous materials, structural quality for pressure retention applications	$\dfrac{0.92}{4} S_T$	$2/3 \times 0.92\ S_y$	$\dfrac{0.92}{4} S_T$	$\dfrac{(1.1 \times 0.92)}{4} S_T R_T$	$2/3 \times 0.92\ S_y$	$2/3 \times 0.92\ S_y R_y$	NA	NA	NA
Ferrous materials, structural quality for nonpressure retention functions	$\dfrac{S_T}{4}$	$2/3\ S_y$	$\dfrac{1.1}{4} S_T$	$2/3\ S_y$	$2/3\ S_y R_y$	NA	NA	NA

Note:
(1) For specific nonferrous alloys and for austenitic materials, two sets of allowable stress values are provided. One set of stresses does not exceed 66.66% of the yield strength at temperature while the other set does not exceed 90% of the yield strength at temperature. The higher stress values should be used only where slightly higher deformation is not in itself objectionable. These higher stress values are not recommended for the design of flanges or other strain sensitive applications.

Nomenclature

NA = not applicable
R_T = ratio of the average temperature-dependent trend curve value of tensile strength to the room temperature tensile strength
R_Y = ratio of the average temperature-dependent trend curve value of yield strength to the room temperature yield strength
S_{Ravg} = average stress to cause rupture at the end of 100 000 h
S_{Rmin} = minimum stress to cause rupture at the end of 100 000 h
S_c = average stress to produce a creep rate of 0.01%/1000 h
S_T = specified minimum tensile strength at room temperature, ksi
S_Y = specified minimum yield strength at room temperature

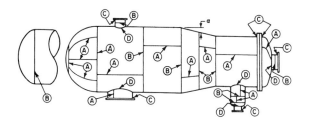

**Fig 4.4 Welded joint categories (ASME VIII-1).
(Source: ASME)**

Table 4.8 Cylindrical shells: basic equations ASME VIII–1

Thin cylindrical shells under internal pressure

For circumferential stress:
 $t = PR/(SE - 0.6P)$
 when
 $t < 0.5R$ or $P < 0.385 \, SE$

For longitudinal stress:
 $t = PR/(2SE + 0.4P)$
 when
 $t < 0.5R$ or $P < 1.25 \, SE$
where
 P = internal pressure
 R = internal radius
 S = allowable stress in the material
 E = joint efficiency factor
 t = thickness of shell material

Note how these equations are expressed in terms of the inside radius of the cylinder shell.

Table 4.8 Cont.

Thick cylindrical shells under internal pressure

For circumferential stress:
$t = R(Z^{1/2} - 1)$
where
$Z = (SE + P)/(SE - P)$

For longitudinal stress:
$t = R(Y^{1/2} - 1)$ for
$t > 0.5R$ or $P > 1.25\ SE$
where
$Y = (P/SE) + 1$

For vessels in ASME VIII Div 2, different equations have to be used.

Axial compression

Vessels can be subject to compressive axial stresses by wind/dead loads and nozzle loads. The general design principle used in ASME VIII is that the acceptable axial load on thick cylinders (with $t/R > 0.5$) is governed by tensile stress limitation, and that on thin cylinders is governed by compressive stress limitation on the shell.

The ASME methodology for calculating allowable axial compressive stress in a thin cylinder shell is (see VIII–1 UG–23):

- Calculate the strain $\in$ from
  $\in\ = 0.125/(R_o/t)$

where

R_o = outside radius of cylinder

t = thickness of cylinder

- Read off the allowable stress corresponding to $\in$ from the stress–strain curves in the code. Note that these curves relate to specific materials and design temperature conditions.

External pressure

The general effect of external pressure on cylindrical vessels is to cause compressive stresses that act to produce buckling. Because of the complexity of the resulting equations, ASME VIII simplifies the situation for design purposes. Figure 4.5 shows the basic methodology.

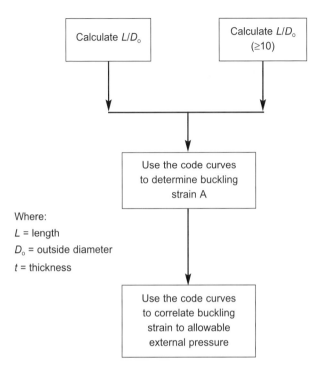

**Fig. 4.5 External pressure on cylinders – basic methodology.
See VIII–1 UG–28 for the full methodology**

Spherical shells and hemispherical dished ends

ASME VIII–1 uses a similar approach to calculate design stresses for internal pressure on spherical shells and the inside of hemispherical 'dished' ends (or *heads*). There are also similar approaches used for external pressure on these components. Table 4.9 gives a summary. Table 4.10 shows comparable information for ellipsoidal dished ends and Table 4.11 gives information for torispherical dished ends.

4.6.4 Flat plates, covers, and flanges

Most plates, covers, and flanges used in pressure vessels are circular and unstayed with a uniform pressure loading over their flat surface. The design rules for these are similar for ASME VIII Div 1 and Div 2. There are two sets of rules governing openings: one set is applied when the opening size is below certain size limits, and the other when it is above. Table 4.12 gives a 'quick reference' summary.

Table 4.9 Spherical shells and hemispherical dished ends

Internal pressure *1 VIII–1 UG–32	External pressure VIII–2 UG–28d
$t = PR/(2SE-0.2P)$ when $t < 0.356R$ or $P < 0.665SE$ when $t/R > 0.356$ the code uses $t = R(Y^{1/3} – 1)$ where $Y = 2(SE+P)/(2SE-P)$ P = internal pressure P_a = allowable external pressure R = internal radius S = stress in material t = thickness of material E = joint efficiency factor A = strain ($\in$) B = allowable compressive stress	The methodology is • calculate the quantity A from $A = 0.125/(R_o/t)$ • Use the code stress–strain diagram to get the corresponding B value. • Calculate the allowable external pressure using $P_a = B/(R_o/t)$
*1 Note this does not take account of liquid weight – see API 620 for this	Note that there are some situations for which modulus E has to be brought into the equation. (See VIII–1 UG–28.)

Table 4.10 Ellipsoidal dished ends

Internal pressure	External pressure
For $D/2h = 2$	The methodology (see VIII–1 UG–33) is:
$t = PD/(2SE\text{-}0.2P)$	for $D/2h = 2$
[See VIII–1 UG–32 (d)]	$t = 1.67P/(2S\text{-}0.2P)$
For other $D/2h$ ratios between 1.0 and 3.0	For other $D/2h$ ratios between 1.0 and 3.0
$t = PDK/(2SE\text{-}0.2P)$ where $K = 1/6\,[2 + (D/2h)^2]$ (See VIII–1 Appendix 1–4)	$t = 1.67PDK(2S\text{-}0.2P)$ An alternative method is • Calculate $A = 0.125/(K_o D_o/t)$ • Determine B from code tables • Calculate allowable pressure from $P_a = B/(K_o D_o/t)$
A = strain D = inside base diameter D_o = outside diameter of ellipsoid P = internal pressure (concave side) S = stress in material t = material thickness E = joint efficiency factor h = depth of dished end K_o = a function of the ratio $D_o/2h_o$ obtained from code tables	Normally it is the greater thickness obtained from the two calculation methods that is used.

Table 4.11 Torispherical 'shallow' dished ends

Internal pressure VIII-1 UG32(e)	External pressure
$t = 0.885\ PL/(SE - 0.1\ P)$	
This is an approximation which assumes that the dished end can be approximated by a spherical radius $L \cong 1.0D$ and knuckle radius $r \cong 0.06D$.	The methodology is the same as for ellipsoidal dished ends except that the term $K_o D_o$ is replaced by the outside crown radius.
For varying spherical and knuckle radii:	
(See VIII–1 Appendix 1–4)	
$t = PLM/(2SE - 0.2P)$	
where	
$M = 1/4\ [3 + (L/r)^{1/2}]$	
and $1.0 \leq L/r \leq 16.67$	
E = joint efficiency factor	
L = inside spherical radius	
P = internal pressure	
S = allowable material stress	
t = thickness of dished end	

Table 4.12 Flat plates, covers, and flanges – basic principles

Thickness (t) of **circular flat plate** is given by (from UG–34):	Thickness (t) of a **non-circular flat plate** is given by:
$t = d\ (CP/SE)^{1/2}$	$t = d\ (ZCP/SE)^{1/2}$
d = effective diameter of plate	Z is a factor used to compensate for lack of uniform membrane support obtained when a plate is circular.
C = coefficient between 1.0 and 0.33 and related to the corner design	
P = design pressure	
S = allowable stress at design temperature	$Z = 3.4 - (2.4d/D) \leq 2.5$
E = welded joint efficiency (depending on the extent of non-destructive testing used)	where D = major dimension of plate d = minor dimension of plate

Table 4.12 Cont.

For **flange rules** see VIII–1 Appendix 2 and VIII–2 Appendix 3

For blind flanges, see VIII–1 Table U–3
and VIII–2 Table AG–150

For rules covering **openings in flat plates**, see VIII–1 UG–39.

As a quick first estimate, for a single opening less than half the plate diameter, the area of reinforcement (A_r) can calculated from:

$A_r = 0.5 d t_r$

where d = diameter of the opening in the flat plate
t_r = minimum required thickness of the flat plate

Note the 0.5 factor; this is inserted to allow for the fact that the stress distribution through the thickness of a flat plate is mainly bending stress (rather than primary membrane stress), if the opening diameter is greater than half the plate diameter. See VIII–1 Appendix 14.

Bolted flanges

Bolted connectors and flanges are in common use on pressure vessels. There are three main types of flange:
- blind flange;
- loose-type flange (flange ring provides the entire strength of the flange);
- integral-type flange.

Basic design rules for all these types are similar and given in VIII–1 Appendix 2. Standard sizes of blind flanges are given in VIII–1 Table U–3. Table 4.12 shows the basic principles. Where a non-standard size is required the basic equation for required thickness, t, is:

for blind flanges: $t = d[(CP/SE) + (1.9 W h_G / SE d^3)]^{1/2}$

(See VIII–1 Table U3)

Similarly, the method of calculating acceptable flange stresses is essentially the same for welding neck flanges as for loose, slip-on, or ring-type flanges.

4.6.5 Vessel openings – general

In principle, openings in the pressure envelope of a vessel generally require the addition of extra material in the vicinity of the opening in order to keep stresses to an acceptable level. The main ways of doing this are:
- increase the wall thickness of the shell;
- add a reinforcing 'doubler' plate around the opening.

ASME VIII–1 gives two methods of evaluating the effect of openings.

The area replacement method

This is sometimes known as the *'reinforced opening method'* and is used when the 'missing' opening area would have provided resistance to primary membrane stresses. This is the only method that can be used for single openings. The principle is that a substitute 'pathway' is required to take the primary membrane stresses. There is a limit on the amount of reinforcement in two directions, i.e. parallel and perpendicular to the vessel shell.

Figure 4.6 shows the situation. Outside this case the code rules are complex, and detailed reference to the code requirements is necessary. Some useful paragraphs are listed below.

Subject	VIII–1 paragraphs
Openings with inherent compensation	UG–36 (c) (3)
Size of openings	UG–36 to UG–42
Reinforcement areas	UW–16.1
Openings in dished ends	UG–37

The ligament efficiency method

This method considers the area of material removed from the vessel shell, compared to the amount left between the holes (i.e. the *ligaments*). The method uses ligament efficiency curves to determine the tensile and shear stresses on various ligament planes. The full equations and curve sets are given in VIII–1 UG–53.5 and UG–53.6.

4.6.6 Heat exchangers

The design of heat exchangers is covered in both ASME VIII–1 and the TEMA (Tubular Exchanger Manufacturers Association) standards. The main areas of interest are:

- rules for tubesheet design   VIII–1 Appendix AA
- design of tube-to-tubesheet welds   VIII–1 UW–20
   and VIII–1 Appendix A
- design rules for expansion joints   VIII–1 Appendix 26

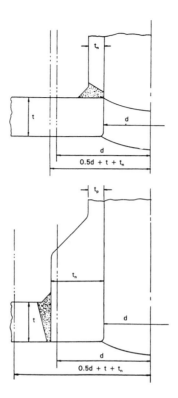

**Fig. 4.6 Reinforcement limits parallel to shell surface.
(Source: ASME)**

- other rules for design, tolerances,   TEMA, 1988
  baffle construction

Types and features

Figures 3.19 and 3.20 show the basic types of TEMA heat exchanger configurations. There are two main design philosophies, split between 'U' tube and fixed-tube exchanger designs. The design of the tubesheet in a 'U' tube exchanger to ASME VIII–1 is based on classical bending theory of a circular plate under pressure. In contrast, fixed tubesheet exchanger design is based on the theory of bending of a circular plate on elastic foundations.

In both cases, the design analysis is complicated and the ASME code makes various assumptions and simplifications.

4.6.7 Special analyses

When detailed design formulae are not given in ASME VIII–1 (normally because the design arrangement is too complex) it is likely that reference will need to be made to VIII–2 (i.e. Div 2). Note that the code is currently being reconsidered, in view of developments in European harmonized standards. A specific purpose of this part of the code is to provide definitions and requirements for stress 'categories', stress concentrations, combinations of stress, and fatigue conditions.

Stress categories

There are three stress categories recognized by ASME VIII–2.

- *Primary stress* – the best examples are longitudinal and circumferential stresses in a cylindrical vessel under internal pressure. Primary stress does not cause load redistribution and can itself be subdivided into:
  – bending stress;
  – general membrane stress (e.g. from internal pressure);
  – local membrane stress (such as caused by local nozzle loads, etc.).
- *Secondary stress* – this is essentially caused by some component of a pressure vessel being *restrained*, either by other components, or by being fixed to something external to the vessel. In contrast to primary stresses (which can cause failure if they rise too high) secondary stresses are self limiting, i.e. they can be redistributed by local yielding, without the vessel having 'failed'. Thermal stresses are classed as having secondary stress characteristics.
- *Peak stress* – this is highly localized stress that, although does not necessarily cause detectable yielding, may result in fatigue. Typical examples are notches and crack-like features. These are classified in VIII–2, as shown in Tables 4.13–4.15.

Applications of Pressure Vessel Codes 93

Table 4.13 Primary stress category. (Source: ASME)

```
                    Primary stress
                   /              \
              Membrane          Bending
              /      \
        General      Local
```

General: A general primary membrane stress is one that is so distributed in the structure that no redistribution of load occurs as a result of yielding. An example is the stress in a circular cylindrical shell due to internal pressure.

Local: Examples of a local primary membrane stress are the membrane stress in a shell produced by external load and the moment at a permanent support or at a nozzle connection.

Bending: An example is the bending stress in the central portion of a flat head due to pressure.

Table 4.14 Structural discontinuity. (Source: ASME)

```
              Structural discontinuity
              /                      \
   Gross structural            Local structural
   discontinuity               discontinuity
```

Gross structural discontinuity: This is a source of stress or strain intensification that affects a relatively large portion of a structure and has a significant effect on the overall stress or strain pattern. Examples of gross structural discontinuities are head-to-shell and flange-to-shell junctions, nozzles, and junctions between shells of different diameters or thicknesses.

Local structural discontinuity: This is a source of stress or strain intensification that affects a relatively small volume of material and does not have a significant effect on the overall stress or strain pattern or on the structure as a whole. Examples of local structural discontinuities are small-nozzle-to-shell junction, local lugs, and platform supports.

Table 4.15 Thermal stress. (Source: ASME)

Thermal stress

This is a self-balancing stress produced by a non-uniform distribution of temperature or by different thermal coefficients of expansion. Thermal stress is developed in a solid body whenever a volume of material is prevented from assuming the size and shape that it normally should under a change in temperature

General

General thermal stress is classified as a secondary stress. Examples of general thermal stress are

1. Stress produced by an axial temperature distribution in a cylindrical shell.
2. Stress produced by a temperature difference between a nozzle and the shell to which it is attached.
3. The equivalent linear stress produced by the radial temperature distribution in a cylindrical shell.

Local

Local thermal stress is associated with almost complete supression of the differential expansion and thus produces no significant distortion. Examples of local thermal stresses are

1. Stress in a small hot spot in a vessel wall.
2. The difference between the actual stress and the equivalent linear stress.
3. Thermal stress in a cladding material.

Analysis of stress combinations

VIII–2 covers various simplified methodologies for calculating acceptable stresses when a component is subjected to a *combination* of stresses. The basic idea is:
- three normal stresses σ_r, σ_l, σ_h are calculated
- three tangential stresses τ_{rl}, τ_{rh}, τ_{lh} are calculated
- the three principal stresses σ_1, σ_2, σ_3 are calculated using

$$\sigma_{\substack{max \\ min}} = \frac{(\sigma_{li} + \sigma_j)}{2} \pm \left[\frac{(\sigma_i - \sigma_j)^2}{4} + \tau_{ij}^2 \right]^{\frac{1}{2}}$$

- the maximum stress intensity, S, as defined in VIII–2 is then assumed to be the largest value obtained from
  $S_{12} = \sigma_1 - \sigma_2$
  $S_{13} = \sigma_1 - \sigma_3$
  $S_{23} = \sigma_2 - \sigma_3$

Fatigue analysis

A large percentage of pressure equipment that fails in service does so because of fatigue. There are many types of fatigue conditions, the most common categories being those induced by cycles of mechanical stress, thermal stress, and externally imposed vibration. ASME VIII–2 covers fatigue conditions in the following paragraphs:
- requirements for a fatigue evaluation: VIII–2: AD-160;
- methodology of fatigue assessment: VIII–2: Appendix 5.

The principle of the fatigue assessment methodology is:
- calculate the maximum stress range (σ_a) for the fatigue cycle – this is the algebraic difference between the maximum and minimum stress intensities in the cycle;
- calculate the alternating stress, which is 50 per cent of the maximum stress range;
- use the fatigue charts in VIII–2 to determine the maximum permissible number of stress cycles for each stress range.

4.6.8 ASME 'intent'

It is often the case that vessels are constructed to 'ASME intent', i.e. without full compliance with the code. Table 4.16 shows a summary of the situation.

Table 4.16 Construction to ASME Intent: Key points

1 Code compliance
- A vessel built to 'ASME intent' will not, by definition, be ASME '*stamped*'. This means that there is no verification of full code compliance and the vessel will *not* be recognized (either by statutory authorities or purchasers) as a vessel that complies fully with the ASME code. Hence the reference to ASME becomes a guideline to engineering practices only. For most vessels built in this way, the adoption of ASME VIII practices is often limited to the outer pressure shell only.

2 Materials
- Vessels built to 'ASME intent' may not use ASME II specified materials – nor will materials have been recertified as required by ASME VIII part UG–10. 'Equivalent' materials should be checked to make sure that elevated- and low-temperature properties are equivalent to the ASTM – referenced materials in ASME II. The most common differences relate to material impact properties.
- Material traceability requirements of ASME are not too difficult to reproduce. A system using EN 10204 certificates of level 3.1B for pressure parts is broadly compliant with the requirements of ASME VIII part UG–94.

3 Manufacturing inspection
- The ASME Authorized Inspector (AI) will normally be replaced with a different third party inspectorate. Although the inspection role will probably be similar to that defined in UG–90, an unauthorized inspectorate cannot authorize 'ASME-stamping' of the vessel.
- NDT is an important area. Most 'ASME-intent' vessels will not use the A, B, C, D joint types specified in ASME VIII and will probably use a simplified extent of NDT (such as RT or UT on seam welds and DP/MPI on nozzle welds). ASME defect-acceptance criteria (UW–51 and 52) may be used but are sometimes replaced by different agreed levels.

4 Documentation
- It it rare that documentation content causes significant problems with 'ASME-intent' vessels. Most competent manufacturers' documentation practices can comply.

4.7 TRD

TRD (Technische Regeln für Dampfkessel Boiler *code of practice for pressure vessels*) is a German standard that cross-references the Arbeitgemeinschaft Druck Behälter (AD Merkblatter) set of technical standards. Table 4.17 shows an outline summary.

Applications of Pressure Vessel Codes 97

Table 4.17 TRD – a summary

Information you need	How to find it in TRD	
1 Responsibilities		
• Certification	TRD 503 specifies that an authorized inspector (AI) shall issue final vessel certification. There are legislative restrictions under the relevant Dampfkessel (steam boiler) regulations controlling which organizations can act as AI for TRD boilers.	
• Manufacturer/purchaser agreement	Such agreements are inferred and can override some parts of the standard –but there is no explicit list of 'agreement items'.	
2 Vessel design	There are four main groups, I, II, III, IV, defined in TRD 600 onwards.	
• Construction categories	Joints are classified by material thickness and type rather than by location in the vessel. The TRD 300 series of documents covers design aspects.	
• Joint types		
3 Materials		
• Permissible materials	TRD 100 gives general principles for materials. In general, DIN materials are specified throughout. Other materials can be used (TRD 100 clause 3.4) but they need full certification by the AI. Typical specified materials are DIN 17102/17155 (plates), DIN 17175/17177 (tubes), DIN 17100/St 37-2/DIN 17243 (forgings), and DIN 17245/17445 (castings). The standards refer to DIN 50049 (EN 10204) material certification levels in all cases.	
4 Manufacture, inspection, and testing	*See TRD section*	**Summary**
• Material identification	TRD 100 (3.4)	Material for pressure parts must be certificated and marked. EN 10204 3.1 A and 3.1 B certificates are required, depending on the material and its application.
• NDT of parent material	HP 5/1 TRD 110/200,	
• Assembly tolerances	TRD 301, TRD 104/201	Visual examination and UT testing is generally required.
• Weld procedures	TRD 201	Circularity (out-of-roundness) and alignment tolerances are closely specified.
• Welder approvals	TRD 201	WPS, PQRs are required for all important welds. Tests are specified.

Table 4.17 Cont.

		Annex 2 specifies welder approval tests to DIN 8560.
• Production test plates	TRD 201	Test plates are required. The number and type of tests to be carried out are closely defined.
• Extent of weld NDT	HP/0, HP5/3	Tables are provided defining NDT requirements for various materials.
• NDT techniques	HP5/3	DIN standards are referenced for all NDT techniques.
• Defect acceptance criteria	HP5/3	HP5/3 gives acceptance levels for RT and UT examinations. Variations can be agreed (Clause 6) by discussions between all parties.
• Pressure testing	TRD 503	A witnessed test is required at 120–150 per cent design pressure, depending on application (see TRD 503, Clause 5).
• Documentation package	TRD 503	A detailed list of documents is not explicitly provided, however, TRD 503 specifies that evidence be provided of material type, heat treatment, and NDT. This infers a full documentation package (similar to BS (PD)5500 and ASME VIII) is required.

4.8 Air receivers

The well-established standard for air receivers was BS 5169: 1992 *Specification for fusion welded air receivers*. This is being replaced by a harmonized European standard, but many air receivers built to the technical guidelines of BS 5169, are still in use and under construction. Table 4.18 and Fig. 4.7 show the key points.

Table 4.18 Air receivers to BS 5169 – key points

1 Code compliance
- BS 5169 requires that 'competent supervision' is required for class I receivers, but does not specify full independent inspection. In *practice* it is common for air receivers to be considered as having equivalent 'statutory status' to BS 5500 vessels, hence they are normally subject to independent survey and certification.
- It is rare to see any type of 'partial code compliance' specified for these air receivers.

2 Materials of construction
- For class I receivers, BS 1449 and BS 4360 plate materials are not permitted.
- Material tensile and bend tests are required.
- Material certificates are mandatory for class I receivers. They are not specified for classes II and III unless the design temperature is outside the range -10 °C to +120 °C.

3 Manufacturing inspection
- WPS/PQR/welder approvals are required for class I and II vessels.
- Most seams are made using double-sided butt welds. If a single-sided weld is used, a backing strip is permitted, subject to satisfactory PQR tests.
- Longitudinal weld test-plates are required for class I and II receivers only.
- For lap welds (allowed for class III head-to-shell joints), the pre-welding fit-up is important. Check that the plates have a tight 'sliding' fit.
- Hydrostatic test pressure is 125 per cent design pressure (P) for class I and 150 per cent P for class II.

4 Documentation
  Although a formal 'Form X' type certificate is not mandatory, BS 5169 does require the manufacturer to issue a 'Certificate of Construction and Test'.

Class definitions		
Class	Definition	Design points
I	Unlimited size and design pressure (P)	Flat end-plates unacceptable. Test plates required for longitudinal seams.
II	P≤35 barG maximum and P (barG) × i.d (mm) ≤37 000	Test plates required for longitudinal seams – but only transverse tensile and bend tests.
III	P≤17.5 barG maximum and P(barG) ≤8 800	Test plates not required.

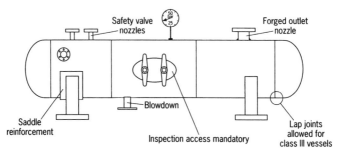

Materials
- Class I receivers use grades of steel from: BS EN 10 207
  BS 1501 (plate)
  BS 1503 (forgings)
  BS 1502/970 (bar)
- Class III receivers can also use plate grades BS 1449 and BS 4360
- Impact tests are required if design temperature is < 0 °C (as in BS 5500 App.D)

NDT requirements				
Location	NDT required			Acceptance criteria
	Class I	Class II	Class III	
Parent plate	Not specified	None	None	Unacceptable defects are:
Circumferential and longitudinal welds	100% RT	None	None	• cracks or l.o.f • elongated slag inclusions (see BS 5169)
Nozzle welds and other welds	Not specified	None	None	• total porosity >6 mm^2 per 25 mm wall thickness in any 645 mm^2 of weld area

Fig. 4.7 Air receivers to BS 5169 (classes I, II, III)

4.9 Shell boilers: BS 2790 and EN 12953

BS 2790: 1992 *Specification for design and manufacture of shell boilers of welded construction* was the existing UK national standard. This will be replaced by the new European harmonized standard EN 12953, with BS 2790 being retained as a technical code of practice. There are still many boilers in use, and under construction, to BS 2790. Figures 4.8 and 4.9 show typical shell boiler layouts and Table 4.19 and Fig. 4.10 show key points from the BS 2790 requirements.

Table 4.20 outlines the separately published parts of EN 12953 due for publishing starting in late 2000.

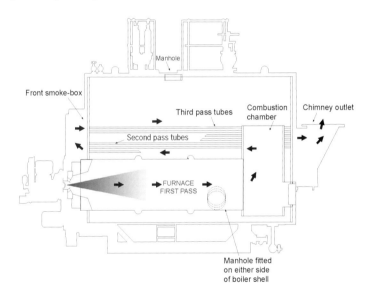

Fig. 4.8 Shell boiler – general layout

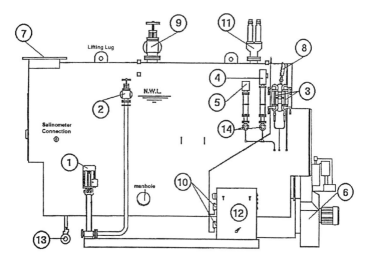

1 Boiler feed pump
2 Feed check valve
3 Water gauge glass assembly
4 Dual control
5 Over-riding control
6 Burner and fan
7 Flue gas exit
8 Pressure gauge
9 Stop valve
10 Pressure switches
11 Safety valve
12 Control panel
13 Blowdown valve
14 Drain valves

Fig. 4.9 Shell boiler fittings

Table 4.19 Shell boilers to BS 2790*: 1992 – Key points

Scope of code BS 2790 covers shell (fire-tube) cylindrical boilers intended for land use to produce steam or hot water. Boilers covered by this standard are vertical or horizontal. The boilers are fabricated from carbon or carbon–manganese steels by fusion welding to class 1, class 2, or class 3 limitations. This standard is not applicable to water-tube boilers and locomotive type boilers.

1 Boiler classifications

Class 1 boiler if either or both apply:

- design pressure > 0.725 N/mm^2 (105 psi)
- design pressure (N/mm^2) x mean diameter (mm) of boiler shell > 920

Table 4.19 Cont.

Class 2 boiler:
- 0.38 N/mm^2 (55 psi) < design pressure <0.725 N/mm^2 (105 psi)
- design pressure (N/mm^2) x mean diameter (mm) of boiler shell <920

Class 3 boiler:
- design pressure <0.38 N/mm^2 (55 psi)
- design pressure (N/mm^2) x mean diameter (mm) of boiler shell <480

Included in the standard are requirements for the installation of safety valves, fittings, and automatic control equipment.

Excluded from the standard are superheaters, economizers, air pre-heaters, ancillary equipment, brickwork, etc.

2 Code compliance
- BS 2790 states that independent survey and certification is required.
- Later additions to the standard require that manufacturers must have an operating quality system in place.
- It is rare to see any kind of 'partial code compliance' specified – although class III vessels have a very low level of verification of integrity.

3 Materials of construction
- These shell boilers are made predominantly from carbon or carbon–manganese steels such as BS EN 10028/BS 1501 (plate) and BS 1503 for forged components. Other similar steels with < 0.25 per cent carbon can be used under the provisions of the standard.
- Test plates are required. Those representing single-sided, full-penetration butt joints require extra root-bend tests.
- Pressure-part material must be identified, but the standard does not explicitly require 'full traceability'.

4 Manufacturing inspection
- WPS/PQR and welder approvals (EN 287, 288) are required.
- Full penetration welds must be used for longitudinal and circumferential seams. Remember the importance of eliminating root defects in this type of weld.
- Longitudinal joints have an alignment tolerance of t/10, to a maximum of 5 mm.
- Weld repairs are allowed – the same acceptance criteria are used.
- There are three sets of defect acceptance criteria quoted in the standard. They are similar to those in BS 5500.
- Make sure that attachment welds do not cross seam or nozzle welds.
- Hydrostatic tests are performed at 150 per cent design pressure for 30 min.

5 Documentation
- There is no mandatory certificate format. The manufacturer must issue a certificate to be authorized by the independent inspection organization.
- A full document package (similar to that for PD 5500) is normally provided.

* Formally replaced by EN 12953

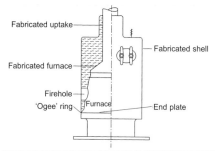

Class definitions

Class	Definition	
I	Design pressure (P) > 7.2 bar and P/10 × diameter (d) mm > 920	There are not a large number of construction differences between class I, II, and III vessels. The main difference is in the NDT requirements
II	Neither of the class I definitions apply	
III	$P \leq 3.8$ bar and P (bar)/10 × (d) ≤ 480	

NDT requirements

Location	NDT requirements			Acceptance criteria
	Class I	Class II	Class III	
Parent end plate	100% UT	100% UT	Not required	BS 5996: Q grade B4 and E3
Other parent plate	Edges only	Edges only	Visual exam only	BS 5996: Q grade E3 (edges)
Longitudinal butt welds	100% RT or UT	10% RT or UT	Visual exam only	BS 2790 gives 3 sets of
Circ^L welds shell to end-plates	100% RT or UT	100% RT or UT	Visual exam only	acceptance criteria for:
Other circ^L welds	10% RT or UT	10% RT or UT	Visual exam only	• Visual examination (see below)
T-butt furnace welds	25% RT or UT	25% RT or UT	Visual exam only	• Radiographic examination
Attachment welds	25% DP or MPI	10% DP or MPI	Visual exam only	• Ultrasonic examination

BS 2790 has its own visual acceptance criteria for welds. They are:
- undercut ≤ 0.5 mm
- shrinkage ≤ 1.5 mm
- overlap not permitted
- excess penetration ≤ 3 mm
- weld reinforcement must be smoothly 'blended in'.

Fig. 4.10 Shell boiler to BS 2790 (classes I, II, III)

Table 4.20 Outline of EN 12953 European harmonized standard on shell boilers

EN 12953 Part	Comments
1 General	Describes the general scope and types of boilers covered.
2 Materials	Lists acceptable EN materials and the procedures for using alternative materials.
3 Design	Explains design materials – with some significant technical changes from BS 2790.
4 Workmanship and construction	Similar content to BS 2790, but with changes in material stress relief requirements.
5 Inspection, documentation, and marking	Contains inspection requirements during manufacture and coordination of activities of the *Responsible Authority* under the PED modules.
6 Requirements for equipment	Covers equipment such as level alarms, pressure limits, etc. fitted to boilers. More comprehensive than BS 2790.
7 Requirements for firing systems	Covers various parts of the combustion system, e.g. fuel handling (but excluding burners).
8 Requirements for safeguards against excessive pressure	Covers the requirements of boiler safety valves to comply with (other) relevant EN standards.
9 Requirements for control and limiting devices and safety circuits	Supports Part 6 of the standard.
10 Requirement for feedwater and boiler quality	Covers a similar area to the existing standard BS 2486.
11 Acceptance tests	Replaces BS 845, covering acceptance tests carried out on new boilers.
12 Requirements for firing systems for boiler solid fuels	Covers solid fuels such as wood.
13 Special requirements for stainless steel sterilizer boilers	An extension to the scope of shell boilers which are generally constructed of carbon–manganese steel.

4.10 Canadian standards association B51-97, part 1 boiler, pressure vessel, and piping code – 1997

Code summary

Scope of code

The accepted scope of this code is limited to boilers (including boiler external piping) and pressure vessels, as defined in Part 1. However, piping, fittings, and the following vessels are excluded from acceptance by The National Board in the USA:
- portable tanks for liquefied petroleum service;
- liquefied natural gas containers;
- compressed natural gas containers;
- pressure vessels for highway transportation of dangerous goods.

This Standard is divided into three parts:
- Part 1 – Boiler, pressure vessel, and pressure piping code;
- Part 2 – High pressure cylinders for the on-board storage of natural gas as a fuel for automotive vehicles;
- Part 3 – Requirements for CNG refuelling station pressure piping systems and ground storage vessels.

Part 1 of the Standard covers all boilers, pressure vessels, fittings, and piping, including blow-off vessels, thermal fluid heaters, hot water, hydropneumatic, and cushion tanks, anhydrous ammonia tanks, and liquefied petroleum gas tanks. It divides pressure equipment into three service categories: benign liquid (like water), non-lethal, and lethal. The following conditions must be satisfied for the equipment to be within the scope of B51.

Benign liquid service [Part 1, Fig. 1(a)]

15 psig $< P <$ 600 psig, $T >$ 150 °F, $V >$ 1.5 cu ft, & $D >$ 6 in
or
$P >$ 600 psig, $T >$ 150 °F, & $D >$ 6 in

Non-lethal service [Part 1, Fig. 1(b)]

15 psig $< P <$ 600 psig, $V >$ 1.5 cu ft, & $D >$ 6 in
or
$P >$ 600 psig, $V >$ 1.5 cu ft, & $D >$ 6 in
or
$P >$ 600 psig & $V >$ 1.5 cu ft

Lethal service [Part 1, Fig. 1(c)]

$P > 15$ psig, $V > 1.5$ cu ft, & $D > 6$ in
or
$P > 15$ psig & $V > 1.5$ cu ft

4.11 CODAP – unfired pressure vessels

Syndicat National de la Chaudronnerie, de la Tôlerie et de la Tuyauterie Industrielle (SNCT)

and

Association francaise des Ingénieurs en Appareils à Pression (AFIAP)

French Code for Construction of Unfired Pressure Vessel (CODAP: 1995)

Code summary

Scope of code

CODAP 95 is applicable to metallic vessels with a minimum internal design pressure of 0.5 bar (7.25 psi) or an external design pressure of 0.1 bar (1.45 psi). Code physical boundaries include nozzle flanges and threaded or welded connections with termination at the first circumferential weld for piping connections. Included within the scope of the code are support components, fasteners, or other devices welded to the pressure boundary. Excluded from the code are fired vessels, multilayer vessels, hooped vessels, and nuclear equipment.

4.12 Water tube boilers: BS 1113/pr EN 12952

BS 1113: 1998

Specification for design and manufacture of water-tube steam generating plant (including superheaters, reheaters, and steel tube economizers)

Code summary

Scope of code

BS 1113: 1998 covers water-tube steam boilers including: economizers, superheaters, steam reheaters, and parts not isolated from the boiler by a stop valve. Requirements for specific valves, gauges and fittings, such as safety valves, water level gauges, feedwater valves, etc. are included. The standard is applicable to natural circulation, forced, assisted, or controlled circulation and once-through boilers. The code boundary extends up to, and including, the valves separating the boiler and superheater from steam pipes to other equipment, water supply pipes, drain pipes, and pipes to the surrounding atmosphere. Excluded from the standard are: brickwork, settings and supports, insulation, air preheaters, mechanical stokers, fuel burning, and ash disposal equipment, and other accessories.

Draft CEN/TC269 draft standard pr EN 12952 water-tube boilers

The draft European harmonized standard for water-tube boilers is pr EN 12952. It will provide a presumption of conformity under the PED. Table 4.21 shows the outline of the contents of the draft standard.

Table 4.21 Outline of pr EN 12952 – draft European harmonized standard – water-tube boilers

pr EN 12952 Part	Comments
1 General	Describes the overall scope of the standard and provides guidance about PED conformity assessment procedures.
2 Materials	Outlines acceptable EN materials, European Approved Materials (EAMs) and Particular Material Appraisals (PMAs).
3 Design	Outlines design methods (not dissimilar to earlier versions of BS 1113). Specifies factors of safety (to be compared with PED guidance), fatigue, and cyclic loading aspects.
4 Life assessment	Covers life assessment methods.
5 Workmanship	Covers most aspects of manufacturing workmanship including bending operations, heat treatment, welding.
6 Inspection	Specifies the duties of the *responsible authority* (instead of 'Notified Body') and the scope of NDT (and defect acceptance levels) for welds.

Note: Further parts 7–16 are in preparation.

4.13 Materials and referenced standards – quick reference*

Component	PD 5500	ASME VIII	TRD
Plates (shell and heads)	BS 1501: Part 1 Gr 164: Carbon steel Gr 223/224: C/Mn steel	ASTM SA–20 (General requirements)	DIN 17 155 and and DIN 17 102
	BS 1501: Part 2 Gr 620/621: low-alloy steel	ASTM SA–202: Cr–Mn–Si alloy steel	Ferritic steels (see TRD 101)
	BS 1502: Part 3 Gr 304/321: high-alloy steel	ASTM SA–240: Cr–Ni Stainless steel (see ASME II)	
	See also EN 10028		
Forged parts (nozzles and flanges)	BS 1503C – Mn steel BS 1503 Stainless steel (austenitic or martensitic)	ASTM SA–266: Carbon steel ASTM SA–336: Alloy steel ASTM SA–705: Stainless and heat-resistant steel	DIN 17100: grades St 37.2, St 37.3, St 44.2 and 44.3 DIN 17243: High-temperature steels
	See also EN 10222		(see TRD 107)
Castings (where used as a pressure-part)	Special agreement required (see Section 2.1.2.3)	ASTM SA–217: Stainless and alloy steels ASTM SA–351:	DIN 17245: grades GS–18 Cr Mo 9–10 and G –X8 Cr Ni 12.
	BS 1504: Carbon, low-alloy or high-alloy steel	Austenitic and duplex steels.	DIN 17445 Austenitic cast steel (see TRD 103)
	See also EN 10222		

Pressure tubes	BS 3604: Parts 1 and 2 Ferritic alloy steel	ASTM SA–192: Carbon steel ASTM SA–213: Ferritic and austenitic alloy tubes (seamless ASTM SA–249: austenitic tubes (welded)	DIN 17175 and 17177: Carbon steel (see TRD 102)
	BS 3605: Part 1 Austenitic stainless steel		
	See also EN 10217 EN 10216		
Boiler/ superheater tubes	BS 3059: Part 1: C–Mn steel	ASTM SA–250: Ferritic alloy	DIN 17066: grades 10 Cr Mo 9–10 and 14 MoV 6–3 for elevated temperature applications (see TRD 102)
	BS 3059: Part 2: Austenitic steel	ASTM SA–209: C–Mn alloy (seamless)	
Sections and bars	BS 1502 Carbon or C–Mn steel	ASTM SA–29: Carbon and alloy steels	General standard steel DIN 17100 – various grades (see TRD 107)
	BS 1502 low-alloy or austenitic stainless steel	ASTM SA–479: Stainless and heat-resistant steels.	

*Note that many British Standards are in the process of being replaced by European harmonized standards. Appendix 3 of this book shows recent status. The CEN/TC reference may be used to find the current state of development.

4.14 Pressure vessel codes – some referenced standards*

Subject	*PD 5500*	*ASME VIII*	*TRD*
Materials			
Plates	BS 1501, 1502 (BS 5996 for NDT). EN 10028	ASTM SA–20, SA –202 (ASTM SA–435 for NDT)	DIN 17155, 17102
Forgings	BS 1503 (BS 4124 for NDT). EN 10222	ASTM SA–266, SA–366	DIN 17100, 17243

Castings	BS 1504 (BS 4080 for NDT) EN 10213	ASTM SA–217, SA–351	DIN 17245, 17445
Tubes	BS 3604, 3605, 3059, 3603, EN 10216, EN 10217	ASTM SA–192, SA–213	DIN 17175, 17177
Materials testing			
Tensile tests	BS EN 10002 (BS 18) BS 3668	ASTM SA–370	DIN 17245, 17445, 50145
Impact tests	BS 131 (BS EN 10045-1)	ASTM E–812	DIN 50115
Hardness tests	BS 240, (BS EN 10003)	ASTM SA–370, E 340	DIN 50103
Chemical analysis	BS 860 Use individual material standards	ASTM SA–751, E354	DIN EN 10036 DIN EN 10188
Welding			
Welding techniques	BS 5135	ASME IX	AD - Merkblatt HP- 5/1,
WPS/PQR/ approvals	BS 4870 (BS EN 288) and BS 4871 (BS EN 287)	ASME IX, ASTM SA–488	HP 5/3 AD HP 2/1 DIN 8563 TRD 201 DIN 8560
NDT			
Radiographic techniques Image quality Ultrasonic techniques	BS 2600, 2910 BS 3971 (BS EN 462–1) BS 3923, 4080 BS 6443	ASTM E94, E1032 ASTM E142, ASTM SA–609, SA–745, E273 ASTM E165, E433 ASTM E709,	AD HP 5/3, DIN 54111 DIN 54109 DIN 54125, 54126 DIN 54 152
Dye-penetrant tests Magnetic particle tests	BS 6072	E1444, SA–275	DIN 54 130
Destructive tests	BS 4870 (BS EN 288), BS 709	ASTM SA–370	HP5/2, HP 2/1
Pressure testing	BS 3636 (gas tightness only)		TRD 503

*Note that many British Standards are in the process of being replaced by European harmonized standards. Appendix 3 of this book shows current status. The CEN/TC reference may be used to find the current state of development.

CHAPTER 5

Manufacture, QA, Inspection, and Testing

5.1 Manufacturing methods and processes

Manufacturing methods for pressure equipment fall into two well-defined categories: fabrication (by fusion welding); casting and forging (of various types). The main applications are shown in Table 5.1.

Table 5.1 Pressure equipment – manufacturing methods

Vessels	Fabricated rolled shells with forged nozzles, manways, nozzles, and flanges. Some nozzles may be castings. Dished ends (heads) are spun – a type of cold deformation process.
Valves	Most general industrial equipment valves have cast bodies. Higher pressure valves are often forged (large diameter offshore pipeline globe valves are a good example).
Gas cylinders	Mostly manufactured by fabrication or one of the forging processes.
Turbines	Sand casting with forged rotors and investment-cast blades.
Pipework	Pipe spools may be either fabricated (rolled then seam-welded) or made by a continuous forging process such as *pilgering*. Low-pressure iron piping can be cast.
Atmospheric tanks	Fabricated from flat plate material.
Tube-type heat exchangers	Cast or fabricated shell and waterboxes. The tube plate is most commonly plate or forged material and the tubes are forged using an *extrusion* process.

Table 5.1 Cont.

Plate-type heat exchangers	Cold-pressed Ti or Cu/Ni plates with heavy plate or forged end pieces.
Boilers	Boiler drums are fabricated from rolled plate and spun heads. Smaller diameter headers for superheaters, economizers, deaerators, etc. are generally forged (effectively as a thick-walled tube) and then smaller bore nozzles welded in. Tubes are generally forged (extruded). Packaged 'shell' boilers have similar manufacturing methods to shell and tube heat exchangers.

5.2 Vessel visual and dimensional examinations

Visual and dimensional examinations are part of the final manufacturing inspection activities carried out on a pressure vessel. Final inspection is mandatory for vessels that are subject to the previous system of statutory certification; it is a normal contractual witness point and is mandatory for various modules under the Pressure Equipment Directive (PED) see Chapter 11. The major points are outlined in Sections 5.2.1 and 5.2.2.

5.2.1 The vessel visual examination

The purpose of the visual examination is to look for problems that are likely to affect integrity.

The vessel exterior

Plate courses	Check the layout of the plate courses against the original approved design drawings.
Plate condition	Check for dents and physical damage. Look for deep grinding marks or obvious grooves deeper than 10 per cent of plate thickness.
Surface finish	General mill-scale on the surface of the plate is acceptable before shot-blasting. Check for any obvious surface 'rippling' caused by errors during plate rolling.
Reduced thickness	Excessive grinding is unacceptable as it reduces the effective wall thickness. Pay particular attention to the areas around the head-to-shell joint; this area is sometimes heavily ground to try and blend in a poorly aligned seam.

Bulging	Check the whole shell for any bulging. This is mainly caused by 'forcing' the shell or head during tack welding to compensate for a poor head-to-shell fit, or excessive out-of-roundness of the shell.
Nozzle flange orientation	Check that the nozzles have not 'pulled' out of true during fabrication or heat treatment of the vessel. This can cause the nozzle flanges to change their alignment relative to axis of the vessel.
Welding	Make a visual examination of all exterior welding. Check that a double-sided weld has not been replaced with a single-sided one. Watch for rough welding around nozzles, particularly small ones of less than 50 mm diameter. It can be difficult to get a good weld profile in these areas: look for undesirable features such as undercut, incomplete penetration or a weld profile that is too convex.

The vessel interior

Head-to-shell alignment	There should be an even weld-cap all the way around the seam.
Nozzle 'sets'	Check the 'set-through' lengths of nozzles protruding through into the vessel.
Weld seams	Do the same type of visual inspection on the inside weld seams as on the outside. Make sure that any weld spatter has been removed from around the weld area.
Corrosion	Check all inside surfaces for general corrosion. Light surface staining may be caused by the hydrostatic test water and is not a cause for concern. In general, there should be no evidence of mill scale on the inside surfaces. If there is, it suggests the plates have not been properly shot-blasted before fabrication.
Internal fittings	Check that these are all correct and match the drawing. The location of internal fittings is also important; make sure they are in the correct place with respect to the 'handing' of the vessel. Check the fit of the manhole door and any inspection covers.

5.2.2 The vessel dimensional check

The dimensional check

Practically, the dimensional check can be done either before or after the hydrostatic test (see Fig. 5.1). Any strains or distortions that do occur will be small, and difficult to detect by simple measurements methods. Dimensional checking can be done using a steel tape measure, with the use of a long steel straightedge and large inside or outside callipers for some dimensions. Dimensional tolerances for pressure vessels tend to be quite wide. There is a technical standard DIN 8570 that gives general guidance for tolerances on fabricated equipment. The main points are outlined below.

Datum lines	Each vessel should have two datum lines: a longitudinal datum (normally the vessel centreline) and a transverse datum. The transverse datum is normally *not* the circumferential weld line; it is located 50–100 mm in from the seam towards the dished head and is indicated on the vessel by deep centre-punch marks.
Manway location	Check the location of the manway with respect to the longitudinal datum line.
Manway flange face	Check that this flange face is parallel to its indicated plane. A tolerance of + 1 degree is acceptable.
Nozzle location	Check the location of the nozzles in relation to the datum lines.
Nozzle flange faces	Check these by laying a long straightedge on the flange face and then measuring the distance between each end of the straightedge and the vessel shell. Nozzle flange faces should be accurate to within about 0.5 degree from their indicated plane. Check also the dimension from each nozzle flange face to the vessel centreline – a tolerance of + 3 mm is acceptable.
Flange bolt holes	Check the size and pitch circle diameter of bolt holes in the flanges. It is universal practice for bolt holes to straddle the horizontal and vertical centrelines, unless specifically stated otherwise on the drawing.

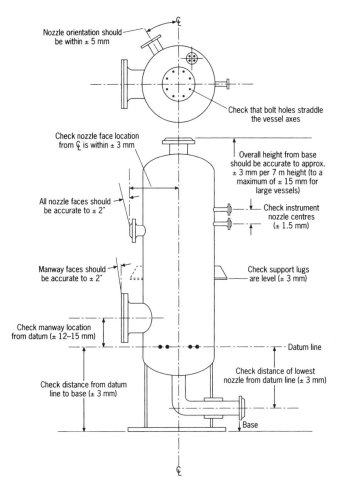

Typical acceptable levels of vessel 'bow'			
Height (mm)	Diameter (mm)		
	<1200	<1300	>1700
<3000 mm	2.5	2	2
3000–9000 mm	4	7.5	6
>9000 mm	5	10	8

Fig. 5.1 Checking vessel dimensions

Vessel 'bow' measurements	The amount of acceptable bow depends on the length (or height) and diameter of the vessel. A small vessel, 3–5 m long with a diameter of up to 1.5 m, should have a bow of less than about 4 mm. A typical power boiler steam drum vessel, approximately 10 m long and 2.5 m diameter, could have a bow of perhaps 6–7 mm and still be acceptable.

5.2.3 Vessel markings

The marking and nameplate details of a pressure vessel are an important source of information. The correct marking of vessel also has statutory implications; it is inherent in the requirements of vessel codes and most safety legislation and directives that the safe conditions of use are clearly indicated on the vessel (see PED pressure equipment marking requirements in Chapter 11). Vessel marking is carried out either by hard-stamping the shell or by using a separate nameplate. Figure 5.2 shows the content and layout of a typical vessel nameplate. For vessels covered by the PED, nameplate marking must include the maximum pressure and minimum temperature. This was not mandatory under previous schemes of vessel certification.

5.3 Misalignment and distortion

Misalignment and distortion can affect the integrity and, therefore, fitness for purpose of the vessel. Misalignment between adjoining plates results in a significant increase in stresses and thermal gradients. This is worst at the shell/head seam where the so-called discontinuity stresses are already high as a result of the constraining effect of the less-flexible dished head (see Chapter 3). For this reason, all the commonly used pressure vessel codes place acceptance limits on misalignment and distortion.

5.3.1 What causes misalignment and distortion?

Misalignment and distortion occur as a result of the practicalities of manufacturing fabricated pressure vessels. The shell plates have to be rolled and then persuaded into alignment with a 'spun' dished head. With thick material, say >20 mm, the material is difficult to roll (it may have to be done hot) and can 'spring back', making accuracy difficult. Conversely, with thin plate, particularly that which is less than 10 mm thick, the main problem is lack of rigidity – it will distort due to heat input during welding.

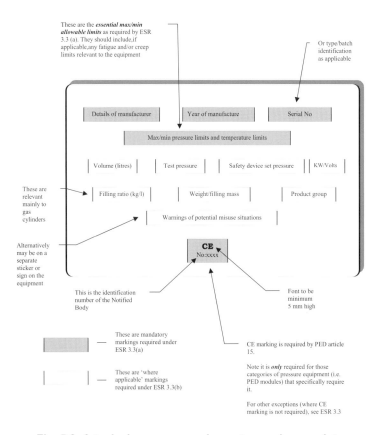

Fig. 5.2 A typical pressure equipment vessel nameplate

5.3.2 Toleranced features

Pressure vessel codes and standards are almost unanimous on which features need to be toleranced to ensure that a vessel retains its fitness for purpose. There are five basic toleranced features:

- **surface alignment** – this really means 'misalignment' of the plate surfaces;
- **straightness**;
- **circumference** – an accuracy measurement;
- **circularity** – also known as *out of roundness* (OOR);
- **profile** – covering dents, bulges, or other departures from a plane or prismatic (three-dimensional) profile.

These five features are used to describe (for inspection purposes) the manufactured accuracy of a pressure vessel. They apply equally to vertical or horizontal vessels and, with a little adaptation, to those comprising more complex shapes such as domes, tapers, or irregular curvature. Figure 5.3 shows a straightforward example of each, and some typical values.

Surface misalignment

Surface misalignment refers to any 'step' that exists between adjacent plates, either head-to-shell or shell-to-shell, after welding is complete. It can occur all the way around a vessel's circumferential seam (if there is a general dimensional mismatch) or, more likely, only on parts of the circumference, as a result of a local mismatch caused when trying to line-up two cylindrical courses of the vessel. *Allowable* surface misalignment is based on a fraction of the plate thickness e – typically $e/4$. There is often a slight difference in the allowable misalignment of longitudinal and circumferential welds. Note how, in Fig. 5.3, the $e/4$ limit is applicable up to a greater parent plate thickness on the circumferential seam than for the longitudinal seam. Longitudinal welds are subject to hoop stress, which is greater than the longitudinal stresses seen by the circumferential welds. Longitudinal seams are, therefore, considered (at least in this context) as 'more critical' and so less tolerant to stress-inducing misalignment.

The same misalignment tolerances apply to the circumferential weld between the dished head and the shell. The spun head is often thicker than the shell plate – this in itself can cause misalignment. As a rough guideline, plate thicknesses used on a commercial pressure vessel will normally vary between about 140 and 170 per cent of the 'minimum acceptable' code thickness requirement, inclusive of any corrosion allowance added.

Straightness

Straightness is expressed as a maximum taper or 'deviation' between parallel planes over unit cylindrical length of the vessel. There is a formal definition in BS 308 Section 10.2. It is easy to measure using wires, straight edges, and a ruler.

Circumference

Circumference is used as a toleranced feature to guard against vessel shells that are 'swaged' as a result of poor rolling or handling. It can also be used as an extra caveat against large differences in the mating dimensions of head and shell (that will result ultimately in surface misalignment as previously described). Figure 5.3 shows typical code values for vessel diameters up to and above 650 mm. These tolerances are the same for head and shell. Thin-walled vessels that exhibit rolling swage-marks are not uncommon – check

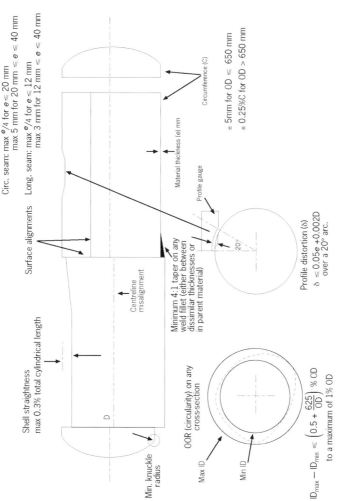

Fig. 5.3 Vessel misalignment and distortions

them using this 'circumference' requirement to see if they are compliant with the code. The easiest way is with a flexible tape that provides a direct reading.

Circularity

Circularity is also known as 'out of roundness' (OOR). A formal definition is provided in BS 308, Clause 10.4. Figure 5.3 shows the basic meaning – it is the difference between the maximum and minimum *internal* diameters on any single cross-sectional plane across the vessel. It is, therefore, a broad measure of ovality or OOR.

Profile

This is a 'catch-all' requirement to put a sensible tolerance on dents, bulges, and general profile variations that could not accurately be classified as out of roundness. Profile tolerance has a strict definition under BS 308, Section 10.7 – it is the *variation* of a surface between two theoretical surfaces separated by a sphere of a diameter equal to the tolerance size. The most common application of profile measurement is to measure bulges and dents by measuring the profile round the vessel in a circumferential direction. The normal way is to use a plywood template cut to the circular profile of the vessel. It should extend at least 20 degrees of arc around the vessel. Some pressure vessel codes allow a greater radial distortion if the size of the dent or bulge is less than one quarter of the distance between the weld seams in the distorted plate. Sharp creases are outlawed by a clause in most vessel standards that require all profile variations to be *gradual*.

5.4 Pressure and leak testing

Virtually all vessels designed to operate above atmospheric pressure will be subject to a pressure test, most often in the manufacturer's works.

5.4.1 The point of a pressure test

The stresses imposed on a vessel during a pressure test are effectively static: they impose principal stresses and their resultant principal strains on the vessel. Hence, a pressure test is the test of the resistance of the vessel only to the principal stress and strain fields, not its resistance to cyclic loadings (that cause fatigue), creep, or the other mechanisms that cause vessels to fail. The pressure test is *not* a full test of whether the vessel will fail as a result of being exposed to its working environment. Hence, a pressure test is not a 'proving test' for vessels that have not been properly checked for defects (particularly weld defects). It is also not a proving test for vessels in which unacceptable defects have been found – so that the vessel can be somehow shown to be fit for purpose, in spite of the defects.

So:

A pressure test is a test for leakage under pressure
and
that is about *all* it is.

5.4.2 The standard hydrostatic test

This is also commonly known as a *hydraulic test*. It is a routine test used when the vessel material thickness and allowable stresses are well defined and there are no significant unknown factors in the mechanical aspects of the design. For single-enclosure vessels (such as drums, headers and air receivers) a single hydrostatic test is all that is required. For heat-exchange vessels such as heaters, coolers and condensers, a separate test is performed on each 'fluid side' of the vessel. General guidelines for hydrostatic tests are given below.

Use these guidelines when witnessing a standard hydrostatic test on any vessel

Vessel configuration

- The test should be done after any stress relief.
- Vessel components such as flexible pipes, diaphragms, and joints that will not stand the pressure test must be removed.
- The ambient temperature must be above 0 °C (preferably 15–20 °C) and above the brittle fracture transition temperature for the vessel material (check the mechanical test data for this).

The test procedure

- Blank off all openings with solid flanges.
- Use the correct nuts and bolts, not 'G' clamps.
- Two pressure gauges, preferably on independent tapping points, should be used.
- It is essential, for safety purposes, to bleed all the air out. Check that the bleed nozzle is really at the highest point and that the bleed valve is closed off progressively during pumping, until all the air has gone
- Pumping should be done slowly (using a low-capacity reciprocating pump) so as not to impose dynamic pressure stresses on the vessel.
- Test pressure is stated in PD 5500, ASME VIII, or the relevant standard. This will not overstress the vessel (unless it is a very special design case). If in doubt use 150 per cent design pressure.
- Isolate the pump and hold the pressure for a minimum of 30 min.

What to look for

- *Leaks*. These can take time to develop. Check particularly around seams and nozzle welds. Dry off any condensation with a compressed air-line, it is possible to miss small leaks if you do not do this. Leaks normally occur from cracks or areas of porosity.
- Watch the gauges for *pressure drop*. Any visible drop is unacceptable.
- Check for *distortion* of flange faces, etc. by taking careful measurements. You are unlikely to be able to measure any general strain of the vessel – it is too small.

5.4.3 Pneumatic testing

Pneumatic testing of pressure vessels is a 'special case' testing procedure used when there is a good reason for preferring it to the standard hydraulic test. Common reasons are:

- refrigeration system vessels are frequently constructed to ASME VIII and pneumatically tested with nitrogen;
- special gas vessels may have an unsupported structure and so are unable to withstand the weight of being filled with water;
- vessels that are used in critical process applications where the process of even minute quantities of water cannot be tolerated.

Pneumatic tests are dangerous. Compressed air or gas contains a large amount of stored energy so, in the unlikely event that the vessel does fail, this energy will be released catastrophically. The vessel will effectively explode, with potentially disastrous consequences. For this reason there are a number of well-defined precautionary measures to be taken before carrying out a pneumatic test on a vessel, and important safety aspects to be considered during the testing activity. These are outlined below along with more general guidelines on witnessing a pneumatic test.

Precautionary measures before a pneumatic test

- PD 5500 requires that a design review be carried out to quantify the factors of safety inherent in the vessel design. NDT requirements are those specified for the relevant Cat 1 or Cat 2 application *plus* 100 per cent surface crack detection (MPI or DP) on all other welds.
- ASME VIII (part UW–50) specifies that all welds near openings and all attachment welds, should be subject to 100 per cent surface crack detection (MPI or DP).
- It is *good practice* to carry out 100 per cent volumetric NDT and surface crack detection of all welding prior to a pneumatic test, even if the vessel code does not specifically require it.

The test procedure

- The vessel should be in a pit, or surrounded by concrete blast walls.
- Ambient temperature should be well above the brittle fracture transition temperature.
- Air can be used, but inert gas (such as nitrogen) is better.
- Pressure should be increased very slowly in steps of 5–10 per cent – allow stablization between each step.
- PD 5500 specifies a maximum test pressure of 150 per cent design pressure.
- ASME specifies a maximum test pressure of 125 per cent design pressure, but consult the code carefully – there are conditions attached.
- When test pressure is reached, isolate the vessel and watch for pressure drops. Remember that the temperature rise caused by the compression can affect the pressure reading (the gas laws).

5.4.4 Vacuum leak testing

Vacuum tests (more correctly termed *vacuum leak* tests) are different to the standard hydrostatic and pneumatic tests. The main applications are for condensers and their associated air ejection plant. This is known as 'coarse vacuum' equipment, designed to operate only down to a pressure of about 1 mm Hg absolute. Most general power and process engineering vacuum plant falls into this category, but there are other industrial and laboratory applications where a much higher 'fine' vacuum is specified .

The objective of a coarse vacuum test is normally as a proving test on the vacuum system rather than just the vessel itself. A vacuum test is much more searching than a hydrostatic test. It will register even the smallest of leaks that would not show during a hydrostatic test, even if a higher test pressure were used. Because of this, the purpose of a vacuum leak test is not to try and verify whether leakage exists, rather it is to determine the *leak rate* from the system and compare it with a specified acceptance level.

Leak rate and its units

The most common test used is the isolation and pressure-drop method. The vessel system is evacuated to the specified coarse vacuum level using a rotary or vapour-type vacuum pump and then isolated.

- The acceptable leak rate is generally expressed in the form of an allowable pressure rise, p. This has been obtained by the designer from consideration of the leak rate in Torr litres/s.

$$\text{Leak rate} = \frac{\mathrm{d}p \times \text{volume of the vessel system}}{\text{time, } t, \text{ (s)}}$$

Note the units are Torr litres per second (Torr ls^{-1}). One Torr can be considered as being effectively equal to 1 mm Hg. Vacuum pressures are traditionally expressed in absolute terms, so a vacuum of 759 mm Hg below atmospheric is shown as +1 mm Hg.
- It is also acceptable to express leak rate in other units (such as l.μm.Hg.s^{-1}, known as a 'lusec') and other combinations. These are mainly used for fine vacuum systems.

5.5 ASME certification

Pressure vessels built to the ASME code have specific inspection and certification requirements.

5.5.1 The role of the AI (Authorized Inspector)

This is a system of third party inspection to ensure that manufacturers of ASME vessels follow the ASME code rules. In the simplest situation, a boiler manufacturer and the boiler purchaser are the first two parties, and an Authorized Inspector (AI) is the independent third party. It is the function of the AI to verify that the manufacturer complies with the code.

Who are they?

An AI is defined by ASME Section I (PG–91) as an inspector employed by a state or municipality of the United States, a Canadian province, or an insurance company authorized to write boiler and pressure vessel insurance. The employer of an AI is called an Authorized Inspection Agency (AIA). In the United States, it has been traditional that the AIA providing authorized inspection for ASME code symbol stamp holders include private insurance companies such as the Hartford Steam Boiler Inspection & Insurance Company, Factory Mutual, or Kemper National.

The AI must be qualified by written examination under the rules of any state of the United States or province of Canada that has adopted the Code. The National Board of Boiler and Pressure Vessel Inspectors (the National Board) grants commissions to those who meet certain qualifications and pass a National Board examination (the website is www.nationalboard.org).

As a condition of obtaining a Certificate of Authorization from ASME to use an ASME code symbol stamp, each Section I (i.e. ASME–I) manufacturer or assembler must have in force a contract with an AIA spelling out the mutual responsibilities of the manufacturer and the AI. The manufacturer, or assembler, is required to arrange for the AI to perform the inspections called for by Section I, and to provide the AI access to all drawings, calculations, specifications, process sheets, repair procedures, records, test results, and any other documents necessary.

ASME certification and its significance

The basic principle is that the vessel manufacturer must certify on a Manufacturers' Data Report Form that all work done complies with the requirements of the ASME Code. In addition, the manufacturer must stamp the boiler with a Code symbol, which signifies that it has been constructed in accordance with the Code. In the case of a complete boiler unit that is not manufactured and assembled by a single manufacturer, the same principles are followed. There is always one manufacturer who must take the overall responsibility for assuring through proper Code certification that all the work complies with the requirements of the Code. An AI must then sign the form as confirmation that the work complied with the applicable Code rules.

5.5.2 Manufacturers' data report forms

ASME Section I has ten different Manufacturers' Data Report Forms (MDRFs) to cover various types of boilers and related components. These forms provide a summary of important information about the boiler (its manufacturer, purchaser, location, and identification numbers) and a concise summary of the construction details used. This includes a list of the various components (drum, heads, headers, tubes, nozzles, openings), their material, size, thickness, type, etc., and other information such as the design and hydrostatic test pressures and maximum design steaming capacity.

The forms and their use are described in ASME Section 1 PG–112 and PG–113, and in Appendix A–350, which contains examples of all the forms and a guide for completing each one. Page A–357 in the Appendix explains which forms should be used in nine different circumstances involving several types of boilers that may be designed, manufactured, and assembled by different stamp holders.

A Master Data Report Form, as the name implies, is the lead document when several different forms are used in combination to document the boiler. The P–2, P–2A, P–3, P–3A, and P–5 forms can be used as Master Data Report Forms. The other forms (the P–4, P–4A, P–4B, P–6, and P–7) usually supplement the information on the Master Data Report Forms and are attached to them. A summary is given below.

The Ten ASME–1 Manufacturers' Data Report Forms

1. **Form P–2**, Manufacturers' Data Report Form for all types of boilers except watertube and electric.

   This form is primarily for fire-tube boilers.

2. **Form P–2A**, Manufacturers' Data Report Form for all types of electric boilers

   This form was specifically designed for electric boilers.

3. **Form P–3**, Manufacturers' Data report Form for water-tube boilers, superheaters, waterwalls, and economizers

   This form can be used to document a complete boiler or major subcomponents.

4. **Form P–3A**, engineering-contractor data report form for a complete boiler unit.

   This is used when an engineering-contractor organization assumes the code responsibility for the boiler normally taken by a boiler manufacturer.

5. **Form P–4**, manufacturers' partial data report form

   This form covers only an individual component part of the boiler such as a drum or header. It merely provides supplementary data for the information shown on the Master Data Report Form.

6. **Form P–4A**, Manufacturers' Data Report for fabricated piping

   PG–112.2.5 explains that this form is used to record all shop or field-welded piping that is within the scope of Section I, but that is not furnished by the boiler manufacturer.

7. **Form P–4B**, Manufacturers' Data Report for field-installed mechanically assembled piping

   This form resembles the P–4A form, but differs from the latter in the certification required.

8. **Form P–5**, summary data report for process steam generators

   This form was devised to cover process steam generators of the waste heat or heat recovery type. These are usually field-assembled arrangements of one or more drums, arrays of heat exchange surface, and associated piping. All components must be certified by their manufacturers on individual data report forms.

9. **Form P–6**, Manufacturers' Data Report supplementary sheet

   This form is used to record additional data when space is insufficient on any other data report form.

10. **Form P–7**, Manufacturers' Data Report for safety valves

    This form lists safety valves, their set pressure and relieving capacity to ensure that safety limits, pressures, etc. are not exceeded.

Completing the data report forms

All ASME–I forms typically provide a place for the manufacturer to certify that the material used to construct the boiler meets the requirements of the Code (no specific edition) and that the boiler design, construction, and workmanship conform to Section I, with the particular edition and Addenda specified. In so certifying, the manufacturer must also identify which Code symbol it is using, and the number and expiration date of the Certificate of Authorization to use it.

The Certificate of Shop Inspection is completed and signed by the AI who has inspected the shop fabrication of the boiler or other components. The field assembly of the boiler is subject to inspection by the assembler's AI, who also completes a certification box giving information about his or her employer, the commission the AI holds, and which parts of the boiler the AI has inspected. The AI must also have witnessed the hydrostatic test of the boiler.

5.5.3 The code symbol stamps

Figure 5.4 and Table 5.2 show details of the ASME code symbol stamps.

The 'S' Stamp

The 'S' stamp covers the design, manufacture, and assembly of power boilers and this is applied to a completed boiler unit. It may also be applied to parts of a boiler, such as a drum, superheater, waterwall, economizer, header, or boiler external piping. Other parts of a boiler such as valves, fittings, and circulating pumps do not require inspection, stamping, and certification when furnished as standard pressure parts under the provisions of PG–11.

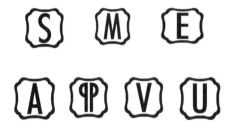

Fig. 5.4 The ASME code symbol stamps

Table 5.2 ASME Section I code symbols stamps

Code symbol stamp	Design capability	Qualified welders & procedures	NDE available	Heat treatment facility available	QC system
S Power boiler	R	R*	R	R*	R
A Boiler assembly	NR	R	R	R	R
PP Pressure piping	R	R	R	R	R
M Miniature boiler	R	R	NR	NR	R
E Electric boiler	R	NR	NR	NR	R
V Safety valve	R	R†	R	R†	R

R = Required

NR = Not required

* It is possible to manufacture certain small boilers without welded joints. In such cases, qualified welders, weld procedures, heat treatment facilities, and non-destructive examination would not be required.

† Not required for valves whose manufacture does not involve welding.

The 'S' stamp may also be used to cover field assembly of a boiler and/or boiler external piping.

ASME generally views the 'S' Stamp as the senior stamp among all the rest, which – in theory at least – allows the holder to engage in any type of Section I construction, i.e. design, manufacture, or assembly, so long as the holder's quality control systems cover that activity.

The 'M' stamp

The 'M' stamp is applied only to miniature boilers, i.e. to boilers that do not exceed the limits specified in PMB–2 of ASME Section I. The miniature classification is thus limited to boilers that are small, have maximum allowable working pressure that does not exceed 100 psi, and do not require radiography or post-weld heat treatment.

The 'E' stamp

The 'E' Stamp is applied to certain electric boilers when the manufacturer of the boiler is not authorized to apply the 'S' or 'M' Stamps, principally because the manufacturer lacks the capability to do Code welding.

The 'A' stamp

The 'A' stamp covers the field assembly of a boiler when carried out by someone other than the manufacturer of the boiler. In such a case, the assembler shares Code responsibility for the completed boiler jointly with the boiler manufacturer, and the assembler's 'A' stamp must supplement the manufacturer's stamp (see PG–107.2 and PG–108.2).

The 'A' stamp may also be applied for the assembly of a boiler external piping.

The 'PP' stamp

The primary application of the pressure piping stamp, PP, is to the fabrication of boiler external piping and to its installation by welding when the piping is not supplied by the boiler manufacturer.

The 'PP' symbol may also be applied to fabrication and shop or field assembly of boiler parts such as superheater, waterwall, or economizer headers, or any construction involving only circumferential welds as covered by PW–41.

The 'V' stamp

The safety valve stamp 'V' may be applied by manufacturers or assemblers of safety valves for power boilers to show that the parts and assembly of the valves comply with Section I requirements.

5.5.4 ASME and the European Pressure Equipment Directive (PED)

Currently, there is no formal agreement in place for mutual recognition between the requirements of the ASME Code and the PED. If such an agreement is not implemented, this means that ASME coded vessels imported into the European Economic Area for use after May 2002 (the end of the PED transition period) will have to be reassessed for compliance with the PED by a European Notified body.

5.6 European inspection terms and bodies: EN 45004: 1995

In the UK and other European countries, organizations providing in-service and new-construction inspection of pressure systems may be (voluntarily) accredited to BS EN 45004: 1995 *General criteria for the operation of various types of bodies performing inspection.* Some important, generally used, inspection definitions are listed below.

Accreditation	Procedure by which an authoritative body gives formal recognition that a body, or person, is competent to carry out specific tasks: BS EN 45020: 1993.
Certification	Procedure by which a third party organization gives written assurance that a product, process, or service conforms to specified requirements: BS EN 45020: 1993.
Inspection	Examination of a product design, product, service, process or plant and determination of their conformity with specific requirements, or on the basis of professional judgement general requirements: BS EN 45004: 1995.
Testing	Measurement of the characteristics of a product by using a technical process.
Conformity assessment	A procedure for checking that a product, service, system, etc. conforms to a standard or specification.
Certificate of conformity	A document (issued under the rules of a certification system) indicating that adequate confidence is provided that a duly identified product, process, or service is in conformity with a specific standard or other normative document: ISO/IEC Guide 2:1991. In the context of independent inspection, it comprises a formal written declaration that the equipment inspected is fit for purpose and conforms to relevant stated requirements.
Notified body	An organization notified to The European Commission by a member state as being competent to undertake conformity assessment of products for CE Marking as a condition of sale in The European Community.
Notifying authority	A government of an EU member state that appoints 'Notified Bodies' from its own country.

Formal definitions and reference literature for the UK is available from SAFed (see Fig. 5.5).

Aim
The aim of the Federation is 'To be the advocate for the independent safety inspection and certification industry on all issues affecting the safety of plant and equipment'.

Objectives
The objectives of the Federation are to advise member companies on safety, certification and legislation relating to manufacturing plant, machinery and equipment and to establish a favourable and public understanding of the industry, through effective communication.

Role
The role of SAFed is 'to add value to the role of its member companies' hence its work is strongly influenced by the role of its member companies, who comprise independent engineering, inspection and non-destruction testing organizations.

Contact details
Mr Richard Morgan: Technical Director, Safety Assessment Federation Limited, Nutmeg House, 60 Gainsford Street, Butlers Wharf, London SE1 2NY. Tel: +44 (0)20 7403 0987 Fax: +44 (0)20 7403 0137.

Fig. 5.5 The Safety Assessment Federation (SAFed)

5.7 The role of ISO 9000

The ISO 9000 series of quality management standards have traditionally been used as a vehicle for controlling the manufacture of pressure equipment and other plant. Established pressure equipment codes have long 'accepted' the role of a quality management standard in controlling the quality of manufactured equipment. The existing structure of ISO 9000 standards are due to be replaced in 2000/2001.

5.7.1 The objectives of the changes

- To make the requirements of the ISO 9000 standards easier to implement in all businesses, not just manufacturing companies.
- To place more emphasis on continual improvement rather than simply compliance with the minimum requirements of an unchanging standard.
- To refocus on the idea of an overall quality management system rather than one which is centred around quality assurance activities.

5.7.2 What will the new standards be?

The final structure and content of the standards is still being developed (ref ISO/TC 176 committee) but the general pattern is likely to be as follows:
- ISO 9001: 2000 *Quality Management Systems* will replace ISO 9001: 1994, ISO 9002: 1994 and ISO 9003: 1994.
- ISO 9002: 1994 and ISO 9003:1994 will be withdrawn as separate standards.
- ISO 9004: 2000 *Quality Management Systems – guidance for performance improvement* will give in-depth detail and explanation of ISO 9001: 2000.
- ISO 19011: 2000 *Guidelines for auditing quality and environmental systems* will replace the existing standard ISO 10011 (formerly BS 7229) covering the techniques of auditing.

It is feasible that these standards will be given additional numbers and titles to fit in with European harmonization activities.

5.7.3 What are the implications?

- *A 'single assessment standard'*. All companies will be assessed against the 'general' quality management standard ISO 9001: 2000. There will no longer be separate standards for companies that are only involved in design, testing, servicing, etc. This means that assessments may need to be *selective*, i.e. only assessing a company on the sections of the standard that are relevant to its activities.
- *Some registration changes.* Companies already certified to ISO 9001: 1994 will probably keep their existing registration number, but those previously certified to ISO 9002: 1994 and ISO 9003: 1994 will need to have their scope redefined.
- *All companies will need to refocus* because of significant change in emphasis of the content of ISO 9001: 2000.

5.7.4 The 'new format' ISO 9001: 2000

The new format is likely to contain sections on:
- *The scope*, and its compatibility with the management systems;
- *Terms and definitions*;
- *Quality Management System (QMS) requirements* – general statements about what a QMS is, and is supposed to do;
- *Management responsibility* – a much expanded version of Clauses 4.1 and 4.2 of ISO 9000: 1994 incorporating aspects such as legal and customer requirements, policy, planning objectives, etc;
- *Resource management* covering staff training (ex Clause 4.18) and new

requirements relating to competence, company infrastructure, and the work environment;
- *Product/service 'realization'* – this section contains the bulk of the content of Clauses 4.3–4.17 of ISO 9000: 1994 but with a lot of changes and restructuring;
- *Measurement, analysis, and improvement* – this incorporates the philosophy of controlling non-conformities and the quality system itself by using reviews and audits (ISO 9001: 1994 Clauses 4.13, 4.14, 4.16, 4.17) but extends the scope to include other, more detailed aspects of measurement and monitoring. It includes greater emphasis on continuous improvement.

Useful references

Information about the status of the 'new-look' ISO 9000: 2000 series of standards can be obtained from BSI (Tel: 00–44–(208) 996 9000, Fax: 00–44–(208) 996 7400, e-mail: info@bsi.org.uk).
Summaries and analysis of ISO 9000 are available on:
- www.isoeasy.org
- www.startfm3.html

Links between ISO 9000 and the PED

There are currently no formal *direct* links between the quality assurance requirements of ISO 9000 and those modules of the PED that require quality assurance monitoring during manufacture (see Chapter 11). Hence manufacturers who are already certified to ISO 9000 do not obtain a presumption of conformity with the requirements of the PED. A separate assessment by a Notified Body is needed.

CHAPTER 6

Flanges, Nozzles, Valves, and Fittings

6.1 Flanges

Vessel flanges are classified by *type* and *rating*. Flange types are shown in Fig. 6.1.

Flanges are rated by pressure (in psi) and temperature, e.g.

ANSI B16.5 classes	
150 psi	
300 psi	
600 psi	Detailed size and design information is given in the ANSI B16.5 standard
900 psi	
1500 psi	
2500 psi	

The type of facing is important when designing a flange. Pressure vessel and piping standards place constraints on the designs that are considered acceptable for various applications (see Fig. 6.2). The corresponding British Standard is BS 4504: 1989.

Table 6.1 shows data for common flange materials.

Table 6.1 Common flange materials

Standard code ASTM	Material	Mechanical requirement				Heat treatment	Hardness HB
		Yield ksi (MPa)	UTS ksi (MPa)	Elongation min(%)	%A min(%)		
A105(N)	–	36.0(250)	70.0(485)	22	30	(Normalized)	187 max
A182	F11	40.0(275)	70.0(485)	20	30		143~207
	F22	45.0(310)	75.0(515)	20	30		156~207
	F304	30.0(205)	75.0(515)	30	50	Solution treatment 1040 °C min	–
	F304L	25.0(170)	70.0(485)	30	50		–
	F316	30.0(205)	75.0(515)	30	50		–
	F316L	25.0(170)	70.0(485)	30	50		–
	F321	30.0(205)	75.0(515)	30	50		–
A266	CL1	30.0(205)	60~85 (415~585)	23	38	Normalized tempered 595 °C min	121~170
	CL2	36.0(250)	70~95 (485~655)	20	33		137~197
A350	LF1	30.0(205)	60~85 (415~585)	25	38	Normalized tempered	197 max
	LF2	36.0(250)	70~95 (485~655)	22	30		
	LF3	37.5(260)	70~95 (485~655)	22	35		
B111 B564	C68700	18.0(125)	50.0(345)			Annealed	
	C70600	15.0(105)	40.0(275)				

Table 6.1 Cont.

Standard Code ASTM	Material	Chemical Composition							
		C	Si	Mn	P	S	Ni	Cr	Mo
A105	–	0.35 max	0.35 max	0.60~1.05	0.040 max	0.050 max	–	–	–
A182	F11	0.10~0.20	0.50~1.00	0.30~0.80	0.040 max	0.040 max	–	1.00~1.50	0.44~0.65
	F22	0.15 max	0.50 max	0.30~0.60	0.040 max	0.040 max	–	2.00~2.50	0.87~1.13
	E304	0.08 max	1.00 max	2.00 max	0.040 max	0.030 max	8.00~11.00	18.00~20.00	–
	F304L	0.035 max	1.00 max	2.00 max	0.040 max	0.030 max	8.00~13.00	18.00~20.00	–
	F316	0.08 max	1.00 max	2.00 max	0.040 max	0.030 max	10.00~14.00	16.00~18.00	2.00~3.00
	F316L	0.035 max	1.00 max	2.00 max	0.040 max	0.030 max	10.00~15.00	16.00~18.00	2.00~3.00
	F321	0.08 max	1.00 max	2.00 max	0.030 max	0.030 max	9.00~12.00	17.00 max	–
A266	CL1	0.35 max	0.15~0.35	0.40~1.05	0.025 max	0.025 max	–	–	–
	CL2	0.35 max	0.15~0.35	0.40~1.05	0.025 max	0.025 max			–
A350	LF1	0.30 max	0.15~0.30	0.60~1.35	0.035 max	0.040 max	0.40 max	0.30 max	0.12 max
	LF2	0.30 max	0.15~0.30	0.60~1.35	0.035 max	0.040 max	0.40 max	0.30 max	0.12 max
	LF3	0.20 max	0.20~0.35	0.90 max	0.035 max	0.040 max	3.3~3.7	0.30 max	0.12 max

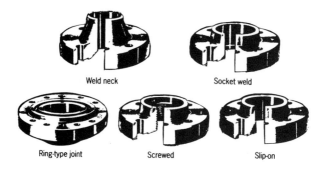

Fig. 6.1 Flange types

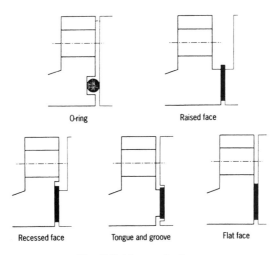

Fig. 6.2 Flange facings

6.2 Valves

Despite the fact that most valves operate under pressure conditions, traditionally they have been considered in a slightly different way to conventional pressure 'vessels'. With the introduction of modern legislation such as the Pressure Equipment Directive (PED) and the Pressure System Safety Regulations (PSSRs), the situation is changing towards one in which valves are formally considered as pressure equipment and are, therefore, bound by similar rules and regulations to those covering vessels

6.2.1 Types of valves

Valve terminology is specified in BS EN 736, Parts 1, 2, and 3.

Figure 6.3 shows common valve types.

6.2.2 Valve technical standards

Technical standards in common use for cast steel valves are:
- valve design and shell wall thickness API 600;
- pressure–temperature rating   ASME/ANSI B16.34;
- face/end dimensions   ASME/ANSI B16.10;
- flange dimensions   ASME/ANSI B16.5 and B16.47;
- inspection and testing   API 598.

The ASME/ANSI standard B16.34 is a valve standard used in conjunction with the ASME codes on boilers and power piping in ASME design installations and, separately, as a valve standard in itself. The standard is relevant to flanged, weld-neck, and threaded end-types for all applications. Theoretically it covers all valves down to the very smallest sizes but, in practice, its main application is for those with a nominal bore of above approximately 60 mm.

The parts of ANSI B16.34 relevant to design, manufacture, and inspection are spread throughout various sections of the document. The key ones are:
- the different valve classes (in Section 2);
- materials and dimensions (Sections 5 and 6);
- NDT scope (Section 8);
- NDT techniques and acceptance criteria (Section 8 and Annex B, C, D, and E);
- defect repairs (Section 8);
- pressure testing.

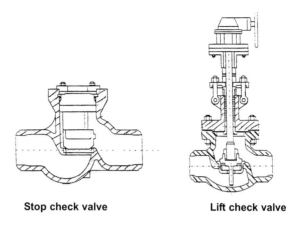

Stop check valve   **Lift check valve**

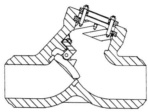

Tilting disc check valve

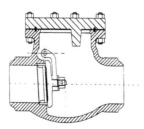

Swing check valve

Fig. 6.3 Some common valve types

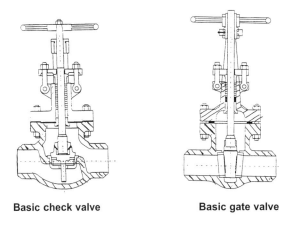

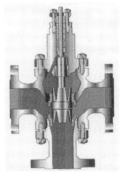

Three-way control valve

Fig. 6.3 Some common valve types (cont.)

Valve classes

ANSI B16.34 divides valves into two main 'classes': standard class and special class. There is a third one called 'limited' class, but it is not used very often. There are a series of pressure–temperature ratings for each type, designated as 150, 300, 400, 600, 900, 1500, 2500, and 4500: these are related predominantly to the valve inside diameter and its *minimum* wall thickness. The higher the number, the larger the wall thickness and the maximum design pressure (see Table 6.2 and Fig. 6.4).

Table 6.2 ANSI B16.34 valves: ratings for Group 1.1 materials. (Source: ANSI B16.34)

A 105 [1,6]	A 515 Gr. 70 [1]	A 675 Gr. 70 [1,4,5]	A 672 Gr. B70 [1]
A 216 Gr. WCB [1]	A 516 Gr. 70 [1,2]	A 696 Gr. C	A 672 Gr. C70 [1]
A 350 Gr. LF2 [1]	A 537 Cl. 1 [3]		

Notes:

[1] Upon prolonged exposure to temperatures above 800 °F, the carbide phase of steel may be converted to graphite. Permissible, but not recommended for prolonged use above 800 °F.

[2] Not to be used over 850 °F.

[3] Not to be used over 700 °F.

[4] Leaded grades should not be used where welded or in any application above 500 °F.

[5] For service temperatures above 850 °F, it is recommended that killed steels containing not less than 0.10 per cent residual silicon be used.

[6] Only killed steel should be used above 850 °F.

Standard class

Temperature °F	Working pressures by classes, (psig)							
	150	300	400	600	900	1500	2500	4500
-20 to 100	285	740	990	1480	2220	3705	6170	11 110
200	260	675	900	1350	2025	3375	5625	10 120
300	230	655	875	1315	1970	3280	5470	9845
400	200	635	845	1270	1900	3170	5280	9505
500	170	600	800	1200	1795	2995	4990	8980
600	140	550	730	1095	1640	2735	4560	8210
650	125	535	715	1075	1610	2685	4475	8055
700	110	535	710	1065	1600	2665	440	7990
750	95	505	670	1010	1510	2520	4200	7560
800	80	410	550	825	1235	2060	3430	6170
850	65	270	355	535	805	1340	2230	4010
900	50	170	230	345	515	860	1430	2570
950	35	105	140	205	310	515	860	1545
1000	20	50	70	105	155	260	430	770

Table 6.2 Cont.

Special class

Temperature (°F)	Working pressures by classes, (psig)							
	150	300	400	600	900	1500	2500	4500
-20 to 100	290	750	1000	1500	2250	3750	6250	11 250
200	290	750	1000	1500	2250	3750	6250	11 250
300	290	750	1000	1500	2250	3750	6250	11 250
400	290	750	1000	1500	2250	3750	6250	11 250
500	290	750	1000	1500	2250	3750	6250	11 250
600	275	715	950	1425	2140	3565	5940	10 690
650	270	700	935	1400	2100	3495	5825	10 485
700	265	695	925	1390	2080	3470	5780	10 405
750	240	630	840	1260	1890	3150	5250	9450
800	200	515	685	1030	1545	2570	4285	7715
850	130	335	445	670	1005	1670	2785	5015
900	85	215	285	430	645	1070	1785	3215
950	50	130	170	260	385	645	1070	1930
1000	25	65	85	130	195	320	535	965

The main difference between the two classes is in the amount of NDT required. 'Special' class valves have to pass a set of NDT tests which are well-defined in the standard (in Section 8). Standard class valves do not have specified NDT, leaving the verification of the integrity of the valve to the pressure test and/or any agreement reached between the purchaser and the manufacturer.

Note:

> Flanged end valves (where the flange is cast integral with the body) cannot be rated as 'special class'.

Valve materials and dimensions

ANSI B 16.34 does not contain any information against which it is possible to check the material selection that has been used for a valve. For this, you need to look at one of the referenced ASME standards, such as ASME B31.3 (Appendix F). There is a materials table in B16.34, but it is simply a summary of the ASME materials with their product forms and 'P' numbers.

There is more information provided about valve *dimensions*. Main areas covered are as follows.
- **Wall thickness** The important criterion is the minimum wall thickness of the valve body, excluding any liners or cartridges that may be fitted. There is useful general information on exactly where on the valve body this minimum wall thickness is measured.
- **Inside diameter** This has to be measured to see if a valve complies with its class designation number, e.g. 150, 300, etc. Section 6.1.2 of the standard explains how this is determined.
- **Neck diameter** Valve body necks also have to meet the minimum wall thickness requirement shown under a valve's class number. There are two slightly different ways to calculate this, depending on whether the valve is rated as 2500 and below, or above 2500.

Scope of NDT

An important part of B16.34 is NDT. Both volumetric and surface NDT scopes are specified for special class valves only. As with some pressure vessel standards, both radiography (RT) and ultrasonic (UT) techniques are acceptable. The areas to be tested are as shown in Fig. 6.5. They concentrate on the weld end, bonnet, neck, and the junction between the body shell and the seat. The exact location and size of the areas to be tested differ between designs and the standard shows the application for several examples. The scope is the same whether radiographic or ultrasonic testing is used.

Specification of surface NDT of special class valves is simpler – the standard requires that all external and accessible internal surfaces of the body and bonnet castings be checked for surface defects. Materials classed under ASME Group 1 can be tested using MPI and DP, while non-magnetic Group 2 and 3 materials must be tested by DP.

Techniques and defect acceptance criteria

Radiography procedures for cast valves are the same as those specified for normal ASME applications, i.e. ASTM E94 and ASTM E142. There is an additional requirement that every radiograph has a penetrameter. The defect acceptance criteria to be used depend on the wall thickness. The assessment of discontinuities is based on the general ASTM reference radiographs for steel castings outlined in the following standards:
- up to 2" wall thickness: ASTM E446;
- 2"–4" wall thickness: ASTM E186.

Practically, most valves have wall thicknesses of less than 2" so ASTM E446 will be the relevant standard to use. Ultrasonic techniques are the

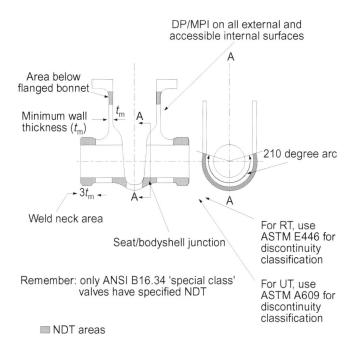

Fig. 6.5 ANSI B16.34 – scope of NDT

Inside diameter (mm)	Class designation 150 300 400 600 900 1500 2500 4500
from 3 mm to about 1270 mm	B16.34 gives values of minimum wall thickness (t_m) required

Three possible additional designations within this table are:
- *Standard class*: no specified NDT requirements
- *Special class*: specified NDT requirements (see B16.34 Section 8 and Annexes B, C, D, and E)
- *Limited class*: for valves <60 mm nominal bore with welded or threaded ends.

Fig. 6.4 ANSI B16.34 valve 'classes'

same as those used under the general standard for ultrasonic testing of castings, i.e. ASTM A609; however, B16.34 does quote its own defect acceptance criteria. These are related to test calibration pieces and are as follows:
- Linear indications are *acceptable* if they are smaller than:
    0.19" (4.8 mm) for sections < 0.5" (12.5 mm) thick;
    0.38" (9.6 mm) for sections 0.5" to 1" (12.5 mm to 25 mm) thick;
    0.75" (19 mm) for sections > 1" (25 mm) thick;

Surface examination techniques are the same as for castings: ASTM E709 for MPI and ASTM E165 for DP. Both have the same acceptance criteria for linear and rounded indications, which are:
- Linear indications:
    0.3" (7.6 mm) long for sections up to 0.5" (12.5 mm) thick;
    0.5" (12.5 mm) long for sections 0.5" (12.5 mm) to 1" (25 mm) thick;
    0.7" (17.8 mm) long for sections over 1" (25 mm) thick.
- Rounded indications:
    0.3" (7.6 mm) diameter for sections up to 0.5" (12.5 mm) thick;
    0.5" (12.5 mm) diameter for sections over 0.5" (12.5 mm) thick.

Defect repairs

In common with many other published standards on pressure equipment castings, B16.34 places *no limit* on the number and extent of defects that can be repaired by welding. The only constraints are those related to the welding technique itself, i.e:
- WPS and PQR to be in accordance with ASME 1X;
- heat treatment to comply with ASME VIII (Div 1);
- repaired areas need to be re-examined using the same NDT technique;
- defect acceptance criteria for radiographed repairs to be in line with ASME VIII Div 1 UW–51.

Hydraulic testing

Table 6.3 shows typical ANSI B16.34 hydraulic test pressures and their duration.

Table 6.3 Valve hydraulic test pressures (ANSI B16.34 basis)

Test by class – all pressures are gauge

Material Group No.	150		300		400		600		900		1500		2500	
	psi	bar	psi	bar	psi	bar	psi	bar	psi	bar	psi	bar	psi	bar
1.1	450	30	1125	77	1500	103	2225	154	3350	230	5575	383	9275	639
1.2	450	30	1125	78	1500	104	2250	156	3375	233	5625	388	9375	647
1.4	375	25	950	64	1250	86	1875	128	2775	192	4650	320	7725	532
1.5	400	28	1050	72	1400	96	2100	144	3150	216	5225	360	8700	599
1.7	450	30	1125	78	1500	104	2250	156	3375	233	5625	388	9375	647
1.9	450	30	1125	78	1500	104	2250	156	3375	233	5625	388	9375	647
1.10	450	30	1125	78	1500	104	2250	156	3375	233	5625	388	9375	647
1.13	450	30	1125	78	1500	104	2250	156	3375	233	5625	388	9375	647
1.14	450	30	1125	78	1500	104	2250	156	3375	233	5625	388	9375	647
2.1	425	29	1100	75	1450	100	2175	149	3250	224	5400	373	9000	621
2.2	425	29	1100	75	1450	100	2175	149	3250	224	5400	373	9000	621
2.3	350	24	900	63	1200	83	1800	125	2700	187	4500	311	7500	517
2.4	425	29	1100	75	1450	100	2175	149	3250	224	5400	373	9000	621
2.5	425	29	1100	75	1450	100	2175	149	3250	224	5400	373	9000	621
2.6	400	27	1025	70	1350	93	2025	140	3025	209	5050	348	8400	580
2.7	400	27	1025	70	1350	93	2025	140	3025	209	5050	348	8400	580

Table 6.3 Cont.

Nominal size DN (NPS)	Minimum duration (min) ≤ PN100 (≤ Class 600)	> PN100 (> Class 600)
≤ 50 mm (2")	1	3
65 mm (2½") – 200 mm (8")	3	8
≥ 250 mm (10")	6	10

Leak testing

Valve leak testing, which is applicable mainly to control valves, is covered by the American Fluid Controls Institute standard FCI 70–2 and API 598. Tables 6.4 and 6.5 show commonly used test data.

Table 6.4 Control valve leakage classes (ANSI/FCI 70-2)

Class	Permittable leakage
I	No leakage permitted
II	0.5% of valve capacity
III	0.1% of valve capacity
IV	0.01% of valve capacity
V	$3 \times 10^{-7} \times D \times \Delta p$ (1/min)

D = seat diameter (mm)
Δp = maximum pressure drop across the valve plug in bar

Table 6.5 Valve leak testing (API 598 basis)

Drops per time unit	Amount (ml)
1 drop/min	0.0625 ml/min
2 drops/min	0.125 ml/min
3 drops/min	0.1875 ml/min
4 drops/min	0.25 ml/min
5 drops/min	0.3125 ml/min
1 drop/2 min	0.03125 ml/min
1 drop/3 min	0.0208 ml/min
2 drops/3 min	0.0417 ml/min

Table 6.5 Cont.

1 drop/4 min	0.0156 ml/min
2 drops/4 min	0.0312 ml/min
3 drops/4 min	0.0469 ml/min
1 drop/5 min	0.0125 ml/min

A rule of thumb conversion is:
  1 ml ≅ 16 drops
  1 drop ≅ 0.0625 ml

6.3 Safety devices

Safety valves are classified under category IV in the PED, except for those that are fitted to vessels in a 'lower' category. Figure 6.6 shows a typical high-pressure safety valve and Table 6.6, typical materials of construction. Safety valve special terminology is given in Section 6.3.2.

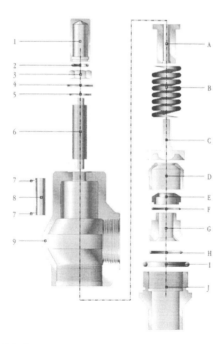

Fig. 6.6 Typical safety relief valve – exploded view

Table 6.6 Typical safety valve materials

No.	Part	Typical material for standard service	Typical material for corrosive service
1	Thread protector	Carbon steel	Carbon steel
2	'O' ring	Rubber compound	Rubber compound
3	Lock nut	Alloy steel	Alloy steel
4	Washer	Carbon steel	Carbon steel
5	Thread seal	Carbon steel	Carbon steel
6	Adjusting screw	Carbon steel – zinc plated	316 stainless steel
7	Drive screw	Stainless steel	Stainless steel
8	Label	Aluminium	Aluminium
9	Body	Carbon steel	Carbon steel
A	Spring keeper	304 stainless steel	304 stainless steel
B	Spring	17.7 stainless steel	Inconel
C	Disc	17.4 stainless steel	316 stainless steel
D	Seat holder	303 stainless steel	316 stainless steel
E	Seat seal	Rubber/plastic compound	Rubber/plastic compound
F	'O' ring	Rubber/plastic compound	Rubber/plastic compound
G	Seat guide	303 stainless steel	316 stainless steel
H	'O' ring	Rubber/plastic compound	Rubber/plastic compound
I	'O' ring	Rubber/plastic compound	Rubber/plastic compound
J	Seat frame	1018 carbon steel	1018 carbon steel

6.3.1 Safety relief valves – principles of operation

All conventional pressure relief valves operate on the principle of system pressure overcoming a spring load, allowing the valve to relieve at a defined capacity.

The basic sequence of operation (see Fig. 6.6) is outlined below.

1. When the valve is closed during normal operation, vessel pressure acting against the seating surfaces is resisted by the spring force. With the system pressure below set pressure by more than 1 or 2 per cent, the valve will be completely leak free.
2. As system pressure is applied to the inlet of the valve, force is exerted on the base of the disc assembly. The force produced by the compression of the spring counters this upward force.

3. When the operating system is below set pressure, the spring housing (or body) of the valve and the outlet are at atmospheric pressure (or at the superimposed back pressure existing in the discharge line.)
4. As operating pressure begins to approach set pressure of the valve, the disc will begin to lift. This will occur within 1 to 2 per cent of set point value and an audible sound will be produced – termed the 'simmer' of the valve.
5. As the disc lifts the gas is transferred from the seat area to the additional area, hence substantially increasing the area being acted on. The result is that the amount of force being applied against the spring compression is dramatically increased. This causes the disc assembly to rapidly accelerate to the lifted position or 'open' condition, resulting in a 'popping' sound.
6. The disc will not stay in the full open position and will begin to drop until an additional pressure buildup occurs. This overpressure condition will maintain the valve in the full open position and allow it to discharge at maximum rated capacity (normally stamped on the valve body in SCFM-standard cubic feet per minute, or equivalent metric units).
7. As the system pressure begins to drop and the spring force overcomes the force created by the disc, the valve will begin to close.
8. The system pressure must drop below the set pressure before the valve will close. This process is termed the the 'blowdown' of the valve.

6.3.2 Terminology – safety valves

Back pressure

Static pressure existing at the outlet of a safety valve device due to pressure in the discharge line.

Blowdown

The difference between actual popping pressure of a safety valve and actual reseating pressure (expressed as a percentage of set pressure).

Bore area

Minimum cross-sectional area of the nozzle.

Bore diameter

Minimum diameter of the nozzle.

Chatter

Rapid reciprocating motion of the moveable parts of a safety valve, in which the disc contacts the seat.

Closing pressure

The value of decreasing inlet static pressure at which the valve disc re-establishes contact with the seat, or at which lift becomes zero.

Disc

The pressure containing moveable member of a safety valve which effects closure.

Inlet size

The nominal pipe size of the inlet of a safety valve, unless otherwise designated.

Leak test pressure

The specified inlet static pressure at which a quantitative seat leakage test is performed in accordance with a standard procedure.

Lift

The actual travel of the disc away from closed position when a valve is relieving.

Lifting device

The device for manually opening a safety valve, by the application of external force to lessen the spring loading which holds the valve closed.

Nozzle/seat bushing

The pressure containing element which constitutes the inlet flow passage and includes the fixed portion of the seat closure.

Outlet size

The nominal pipe size of the outlet passage of a safety valve unless otherwise designated.

Overpressure

The pressure increase over the set pressure of a safety valve, usually expressed as a percentage of set pressure.

Popping pressure

The value of increasing inlet static pressure at which the disc moves in the opening direction at a faster rate as compared with corresponding movement at higher or lower pressures. It applies only to safety or safety relief valves on compressible fluid service.

Pressure-containing member

A pressure-containing member of a safety valve is a part which is in actual contact with the pressure media in the protected vessel.

Pressure retaining member

A pressure retaining member of a safety valve is a part which is stressed due to its function in holding one or more pressure-containing members in position.

Rated lift

The design lift at which a valve attains its rated relieving capacity.

Safety valve

A pressure relief valve actuated by inlet static pressure and characterized by rapid opening or pop action.

Seat

The pressure-containing contact between the fixed and moving portions of the pressure-containing elements of a valve.

Seat diameter

The smallest diameter of contact between the fixed and moving members of the pressure-containing elements of a valve.

Seat tightness pressure

The specific inlet static pressure at which a quantitative seat leakage test is performed in accordance with a standard procedure.

Set pressure

The value of increasing inlet static pressure at which a safety valve displays the operational characteristics as defined under 'popping pressure'. It is often stamped on the safety valve nameplate or body.

Simmer

The audible or visible escape of fluid between the seat and disc at an inlet static pressure below the popping pressure and at no measurable capacity. It applies to safety valves on compressible fluid service.

For further details see ASME PTC 25.3.

6.4 Nozzles

Nozzles are common features on pressure vessels. Most are forged or cast but some larger ones are fabricated from thick plate or tube. The type of nozzle used depends heavily on the nature of the compensation (see Chapter 3) needed to restore the pressure-resisting strength of the vessel shell. Nozzle openings above a certain diameter (specified in individual vessel codes) need the addition of a reinforcing pad to restore shell strength. When strength compensation is needed, nozzles may be '*set-in*' or '*set-through*' the vessel shell. Figure 6.7 shows the most commonly used types.

An alternative type of nozzle used for some applications (typically on thick-walled headers) is the '*weldolet*' (see Fig. 6.8). These are heavily chamfered where they fit into the vessel or header shell and require a large multi-layered weld. In all nozzle fittings, the specification of the weld leg length and throat thickness is critical to the joint strength – in some cases it plays a part in the compensation for loss of strength from the nozzle opening. Figure 9.20 in Chapter 9 shows typical ultrasonic NDT operations used for nozzles.

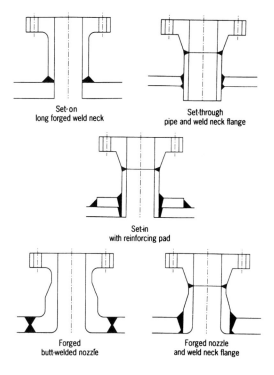

Fig. 6.7 Nozzle types

Fig. 6.8 Nozzle and fitting shapes

6.5 Power piping – ASME/ANSI B31.1 code

The ASME/ANSI code B31.1 *Power piping* parallels the philosophy of ASME I: *Power boilers*. The principles used on allowable stress levels, for example, are similar to those in ASME I, making ANSI B31.1 generally more conservative than some other piping codes. Perhaps, because of this conservatism, ANSI B31.1 is used for power piping in a large proportion of high-pressure boilers worldwide, even where the boiler itself is not constructed to the ASME code.

The scope

Formally, ANSI B31.1 covers external piping (i.e. external to the boiler) only. This is generally taken to mean piping that is positioned after the first flange or welded circumferential pipe joint after the boiler components. Figures 6.9 and 6.10 show two figures taken from Chapter I of the Code relating to drum-type boilers and spray-type desuperheaters.

The contents of ASME/ANSI B31.1 Power piping

Chapter I   **Scope and definition**

Chapter II **Design**
This covers the pressure design of piping components, including various pipe spool shapes, closures, flanges and blanks, reducers, etc. It also covers the selection and limitations of piping components and a wide selection of types of pipe joints.

Chapter III **Materials**
Sets out details of acceptable piping and component materials including cast iron, steel, and some non-ferrous alloys.

Chapter IV **Dimensional requirements**
An important aspect of this chapter is the comparison in dimensional requirements for standard and non-standard piping components.

Chapter V   **Fabrication, assembly, and erection**
This gives extensive requirements on welding operations for pipework.

Chapter VI **Examination, inspection, and testing**
The main areas are: role of the Authorized Inspector (AI), weld examination, and pressure testing (hydraulic and pneumatic). Note that ANSI B31.1 contains its own defect acceptance criteria for welded joints.

Flanges, Nozzles, Valves, and Fittings 159

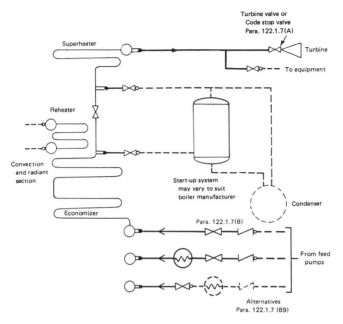

Administrative jurisdiction and technical responsibility

───── Boiler Proper – The ASME Boiler and Pressure Vessel Code (ASME BPVC) has total administrative jurisdiction and technical responsibility. Refer to ASME BPVC Section I Preamble.

●───── Boiler External Piping and Joint (BEP) – The ASME BPVC has total administrative jurisdiction (mandatory certification by Code Symbol stamping, ASME Data Forms, and Authorized Inspection) of BEP. The ASME Section Committee B31.1 has been assigned technical responsibility. Refer to ASME BPVC Section I Preamble, fifth, sixth, and seventh paragraphs and ASME B31.1 Scope, Para. 100.1.2(A). Applicable ASME B31.1 Editions and Addenda are referenced in ASME BPVC Section I, Para. PG-58.3.

○── ── - Nonboiler External Piping and Joint (NBEP) – The ASME Code Committee for Pressure Piping, B31, has total administrative and technical responsibility. See applicable ASME B31 Code.

Code jurisdictional limits for piping – forced flow steam generator with no fixed steam and water line

Fig. 6.9 ANSI B31.1 code limits for piping-steam generation. (Source: ANSI B31.1, Fig. 100.1.2(A))

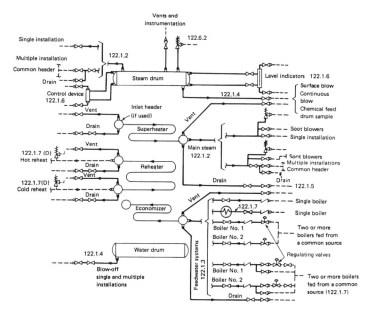

Administrative jurisdiction and technical responsibility

──────── Boiler Proper – The ASME Boiler and Pressure Vessel Code (ASME BPVC) has total administrative jurisdiction and technical responsibility. Refer to ASME BPVC Section I Preamble.

●──────── Boiler External Piping and Joint (BEP) – The ASME BPVC has total administrative jurisdiction (mandatory certification by Code Symbol stamping, ASME Data Forms, and Authorized Inspection) of BEP. The ASME Section Committee B31.1 has been assigned technical responsibility. Refer to ASME BPVC Section I Preamble and ASME B31.1 Scope, Para. 100.1.2(A). Applicable ASME B31.1 Editions and Addenda are referenced in ASME BPVC Section I, Para. PG-58.3.

○── ── ── Nonboiler External Piping and Joint (NBEP) – The ASME Code Committee for Pressure Piping, B31, has total administrative and technical responsibility. See applicable ASME B31 Code.

Fig. 6.10 ANSI B31.1 code limits for piping-drum type boilers. (Source: ANSI B31.1, Fig. 100.1.2(B))

6.6 Fittings

6.6.1 Pressure equipment fittings

Figure 6.11 shows some other common pressure equipment fittings and the common terminology.

6.6.2 Pipework classification

Pipes are normally shown on Process and Instrumentation Diagrams (PIDs) using a system of classification or designation that indicates the pipe material, size and design code. Table 6.7 shows a popular designation system in use.

Elbows

90° long radius elbow   90° short radius elbow   45° long or short radius elbow

The function of an elbow is to change direction or flow in a piping system. They are split into three groups which define the distance over which they change direction, which is expressed as a function of the distance from the centre line of one end to the opposite face. This is known as the centre to face distance and is equivalent to the radius through which the elbow is bent.

Tees and crosses

Reducing tee   Reducing cross   Equal cross   Equal tee

The function of a tee is to permit flow at 90° to the main direction of flow. The main flow passed through the 'run' while the 90° outlet is known as the 'branch'. The equal tee is manufactured with all three outlets being the same size.

Reducers

Eccentric reducer   Concentric reducer   Cone type concentric reducer

The functions of both types of reducer is to reduce from a larger to a smaller pipe size, this obviously results in an increased flow pressure. With the eccentric reducer the smaller outlet end is off centre to the larger end enabling it to line up with one side of the inlet and not with the other.
The concentric reducer is so manufactured that both inlet and outlet ends are on a common centre line.

180° returns

180° long radius return   180° short radius return   End cap

The function of the 180° return is to change direction or flow through 180° and there are two basic types, long radius and short radius. Both types have a centre to centre dimension double the matching 90° elbow. The primary applications for these fittings is in heater coils and heat exchangers.

The function of the end cap is to block off the end of a line in piping system. This is achieved by placing the end cap over the open line and welding around the joint.

Fig. 6.11 Pressure fittings

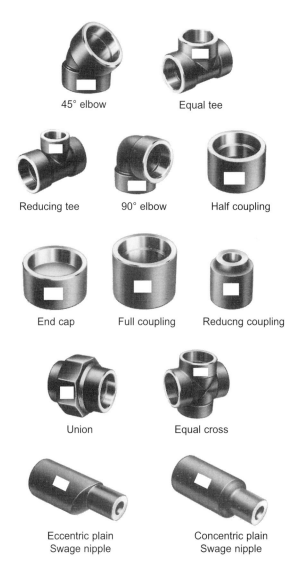

Fig. 6.11 Pressure fittings (cont.)

Table 6.7 Pipeline designation system

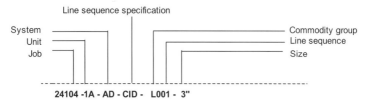

	Line sequence specification	
First character	*Second character*	*Third character*
A Aluminium	1 PN20/Cl 150 (ASME B16.5)	D ASME B31.1
B Copper alloy	2 PN50/Cl 250 (ASME B16.1)	E ASME B31.3
C Carbon steel	3 PN50/Cl 300 (ASME B16.5)	F NFPA
D Ductile iron	4 PNPN68/Cl 400 (ASME B16.5)	G ASME B31.8
E Polyethylene	5 Special rating	H ASME Sec 1
F Fibreglass	6 PN110/Cl 600 (ASME B16.5)	J AWWA
G Galvanized	7 Pressfit rating	N ANSI Z223.1
H Lined cast iron	8 Cl 800 (ANSI B16.1/API 602)	P ASME B31.4
I Cast iron	9 PN150/Cl 900 (ASME B16.5)	Q Plumbing code
J Lined ductile iron	A PN3/Cl 25 (ASME B16.1)	R ASME B31.5
K Alloy steel	B PN4/Cl 50	S ASME B31.9
L Low- temp. carbon steel	C PN5/Cl 75	
M Monel	D 7 barg (manf'rs rating)	
N Nickel alloy	E PN20/Cl 125 (ASME B16.1)	
P PVC	F 12 barg (manf'rs rating)	
Q Concrete	G 14 barg (manf'rs rating)	
R Rubber-lined carbon steel	H 69 barg (manf'rs rating)	

Table 6.7 Cont.

S	Stainless steel	J	PN260/Cl 1500 (ASME B16.5)
T	Titanium	K	139 barg (manf'rs rating)
U	Alloy 20	L	PN420/Cl 2500 (ASME B16.5)
V	Vitrified clay	M	207 barg (manf'rs rating)
W	Epoxy lined carbon steel	N	PN760/Cl 4500 (ASME B16.5)
X	Copper-dosed carbon steel	P	413 barg (manf'rs rating)
Y	Impact grade carbon steel	Q	620 barg (manf'rs rating)
Z	Plastic lined carbon steel	R	689 barg (manf'rs rating)
1	Cast iron; high silicon	S	As specified
2	Cement lined carbon steel	T	As specified
3	PVC	U	<50 mm
4	Alloy steel 1% Cr	V	50 mm–100 mm
5	As specified	W	100 mm–150 mm
6	Alloy steel 5% Cr	X	150 mm–250 mm
7	Polyethylene	Y	>250 mm
8	As specified	Z	Special class 2500 (B16.34)
9	Alloy steel 9% Cr		

CHAPTER 7
Boilers and HRSGs

Boilers are a major type of pressure equipment. There are a wide variety of types, ranging from simple domestic electric boilers to large coal-fired boilers and heat recovery steam generators (HRSGs) used for power station applications. All work using the same fundamental principles.

7.1 Fundamentals of heat transfer

Water can exist in three physical states: solid (as ice), liquid, or gas (vapour) depending on the corresponding temperature and pressure. The principle of steam generation is the conversion from the liquid to the vapour state using the addition of heat. The process is influenced by several fundamental thermodynamic definitions:

7.1.1   Specific heat, c

Specific heat is the quantity of heat required to produce a unit temperature change in a unit mass of water

In SI units:

Specific heat, c, kJ/kgK = amount of heat in kiloJoules required to raise the temperature of 1 kg of water by 1 ºC.

In imperial units:

Specific heat, c, = amount of heat in BTU required to raise the temperature of 1 lb of water by 1 ºF.

Specific heat is often referred to by its alternative name 'thermal capacity' on the basis that it represents the 'capacity' of a fluid to absorb heat.

7.1.2   Enthalpy, h

Enthalpy is a measure of the total stored internal energy of water or steam. Enthalpy changes are a function of temperature and pressure. It is subdivided into several types.

In SI units:
  Enthalpy of saturated liquid $= h_f$ kJ/kg
  Enthalpy of superheated vapour $= h_g$ kJ/kg
In imperial units:
  Enthalpy of saturated liquid $= h_f$ BTU/lb
  Enthalpy of superheated vapour $= h_g$ BTU/lb

The term 'enthalpy' is in common use in both systems of units, but there are slight differences when it is applied to conditions that exist during a change of state from a liquid to a vapour (water to steam).

7.1.3 Latent heat

Latent or 'hidden heat' is heat that, although it is added to a fluid, does not result in a rise in temperature; instead, it results in a change of state (from ice to water or water to steam). There are two relevant types: latent heat of fusion and latent heat of vaporization.

Latent heat of fusion (see Fig. 7.1) is the amount of heat required to melt a unit mass of ice and is rarely relevant to boiler applications.

In SI units:

Latent heat of vaporization h_{fg} in J/kg is the amount of heat required to change 1 kg of liquid water at its boiling point to 1 kg of vapour (steam).

In imperial units:

Latent heat of vaporization h_{fg} is the amount of heat required to change 1 lb of liquid water at its boiling point to 1 lb of vapour (steam).

The term latent heat of condensation refers to the reverse condition where 1 kg(lb) of steam at 100 °C (212 °F) is cooled (heat removed) to form 1 kg(lb) of liquid water at the same temperature.

7.1.4 Steam characteristics

The proportion, by weight, of 'dry' vapour in a steam and water mixture is termed the 'dryness fraction' x. An alternative term is the steam 'quality'. For example, a mixture containing 5 per cent water vapour and 95 per cent steam has a dryness fraction of $x=95$ per cent (or a quality of 95 per cent).

Fig. 7.1 Principle – latent heat of fusion

Figures 7.2 and 7.3 show the basic temperature–pressure and temperature–enthalpy diagrams for water and steam. Figures 7.4 and 7.5 give pressure and temperature conversion factors. Appendix 1 gives 'steam table' properties data.

Saturated steam

Steam is defined as being *saturated* (i.e. saturated with heat) when it contains all the heat it is able to hold at the boiling temperature of water, without turning into steam. Dry saturated steam contains very little moisture and is at saturation temperature, for the pressure at which it exists. Saturation temperature is, therefore, the temperature at which water boils, at a given pressure. For each saturation temperature, there is a corresponding pressure called the 'saturation pressure'.

Superheated steam

Steam is superheated when it is heated above its corresponding saturation temperature, at a given pressure. Under these conditions, the steam contains no moisture and cannot condense until its temperature has been lowered to that of saturated steam at the same pressure. *The degree of superheat* (in °C or °F) is the amount that the steam is heated above the saturation temperature corresponding to the temperature it is at.

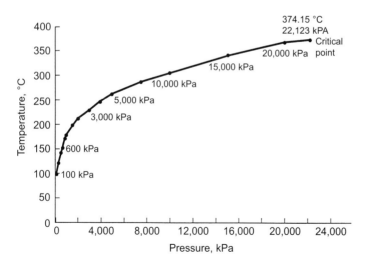

Fig. 7.2 Pressure–temperature relationship for steam (SI units)

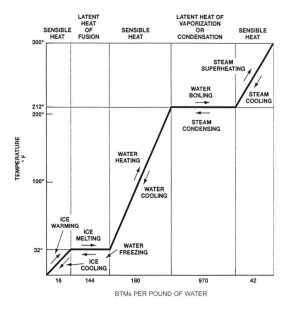

Fig. 7.3 Steam/water temperature – enthalpy diagram (imperial units)

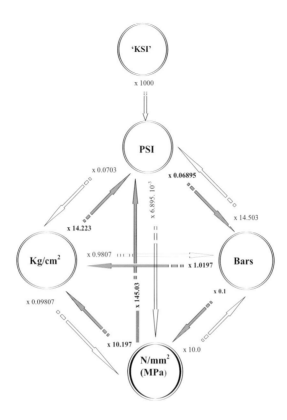

Fig. 7.4 Pressure conversions

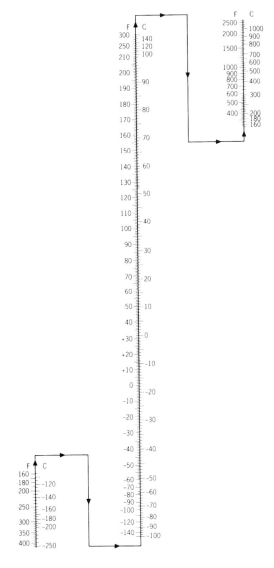

Fig. 7.5 Temperature conversions

The critical point

The 'critical point' is the condition under which the properties of water and steam are identical and so, strictly, the separate terms 'water' and 'steam' no longer apply. The critical point is at conditions of 22123 kPa (3208.2 psi absolute) with a corresponding saturation temperature of 374.15 °C (705.5 °F).

7.1.5 Gas characteristics

Table 7.1 shows typical analyses for different gas streams.

Table 7.1 Gas analyses

Type of gas (generic)	% by volume							
	CO_2	H_2O	N_2	O_2	SO_2	H_2	CH_4	CO
GT exhaust	3	7	75	15	–	–	–	–
Flue gas (natural gas)	8	18	71	3	–	–	–	–
Flue gas (oil)	12	12	73	3	–	–	–	–
Reformed gas	6	36	–	–	–	46	3	9
Sulphur combustion gas	–	–	80	9	11	–	–	–

7.2 Heat recovery steam generators (HRSGs)

7.2.1 General description

Heat recovery steam generators (HRSGs) are the steam-raising component of combined cycle gas turbine (CCGT) power plants and provide a good example of the use of pressure equipment components in a modern plant application. They may have auxiliary firing by gas or oil, but are more often heat recovery units only, deriving the heat for steam-raising from the exhaust gases of the gas turbines. HRSGs generally comprise several interlinked systems designed to operate at different pressures and temperatures. There are many differing proprietary designs. Figures 7.6 and 7.7 show a typical modern, high-efficiency, triple-pressure HRSG design used in CCGT power stations. The basic components are:
- low-pressure(LP) system;
- intermediate-pressure (IP) system;
- high-pressure (HP) system;
- HRSG gas flow path.

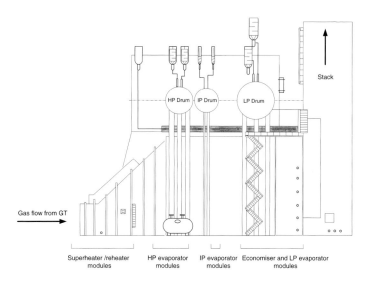

Fig. 7.6 A modern, triple-pressure HRSG

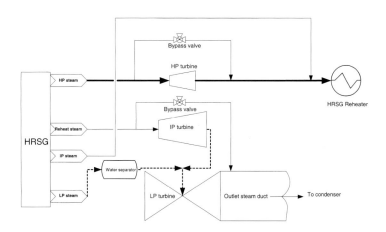

Fig. 7.7 HRSG steam turbine circuits

Low-pressure (LP) system

The LP system operates in a manner similar to that of a natural circulation boiler, consisting of an economizer and LP drum. The drum acts to maintain the equivalent of a 'static head' on the system, while also containing a steam phase at low pressure.

Intermediate-pressure (IP) system

The IP system consists of economizer, evaporator and superheater sections, and an IP drum. The IP feed water gains heat in the IP economizer and evaporator stages before it enters the IP drum. Steam leaves the drum still in saturated condition, before entering the IP superheater stage.

High-pressure (HP) system

The HP system consists of a HP drum, economizer and evaporator sections, and multiple (3 or 4) superheat sections. High-pressure feedwater is supplied to the HP economizer inlet manifold, while steam generated in the evaporator sections flows upwards, via riser tubes, to the HP drum. The drum contains centrifugal steam/water separators that separate out entrained water and return it to the drum.

From the HP drum, steam is directed to various final superheat stages. Most power station HRSGs also have a reheat stage, reheating the steam after partial expansion in a HP turbine, to improve cycle efficiency. Superheat temperature control is achieved by water spray attemperators (often termed 'desuperheaters', see Fig. 7.8). A typical power station boiler will have two stages of desuperheat, one after the final superheater stage (the 'superheat desuperheater') and one controlling the temperature of the steam after the reheat section (the 'reheat desuperheater').

HRSG gas flow path

Exhaust gas from the gas turbine (GT) enters the HRSG and flows sequentially through the sections in (typically) the following order:
- HP superheater 1
- HP reheater 1
- HP superheater 2
- HP reheater 2
- HP superheater 3
- HP superheater 4
- HP evaporator
- IP superheater 1
- HP economizer 1
- HP economizer 2 and 3
- IP evaporator

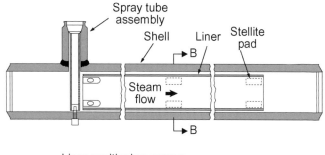
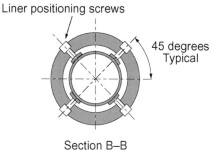

Fig. 7.8 An HRSG water-spray desuperheater

- HP economizer 4
- IP economizer 1
- HP economizer 5
- LP evaporator
- LP economizer

7.2.2 HRSG operation

Table 7.2 shows typical operating conditions for a triple-stage HRSG.

Table 7.2 Typical operating temperatures and pressures: triple-stage power station HRSG

Parameter	Operating condition
LP drum	9 bar, 180 °C
IP drum	35 barg, 253 °C
HP drum	133 barg, 343 °C
Reheater inlet	26 barg, 343 °C
Reheater outlet	25 barg, 545 °C
Gas inlet to HRSG	620 °C
Gas outlet from HRSG (to stack)	84 °C
Circulation ratio	4 to 5
Full-load HP steam flow	80 kg/s @ 103 barg
Full-load IP steam flow	10 kg/s @ 25 barg
Full-load LP steam flow	10 kg/s @ 3.5 barg

Boiler circulation options

Boiler circulation is the mechanism by which water, steam, and mixtures of the two move throughout the boiler circuits. There are two main options: natural and forced.

Natural circulation

In natural circulation HRSGs, the driving force behind the circulation is solely the density difference between fluids in the different parts of the system (Fig. 7.9). There is no circulation pump. Steam/water mixtures in economizer sections of the HRSG have a density of approximately half that of saturated water in the downcomers, thereby encouraging circulation. The differential *de*creases as boiler pressure *in*creases, because the difference between the densities of water and steam, which is the motive force for natural circulation, becomes smaller. Higher pressure HRSG designs are, therefore, less suitable for natural circulation.

Forced circulation

In forced circulation HRSGs, circulation through the boiler circuits is provided by separate mechanical pumps, often located in the downcomer circuit from the steam drums. This is used in higher-pressure designs. An important criterion is the 'circulation ratio'.

> Circulation ratio = mass of water entering downcomers/mass of steam in the steam:water mixture leaving the heating banks.

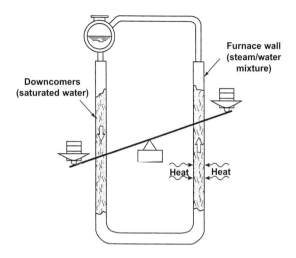

Fig. 7.9 Boiler principles – natural circulation

The circulation ratio is broadly related to boiler pressure and is higher at low boiler loads, as less steam is generated under such conditions.

Safety valves

HRSGs are fitted with safety valves to protect the steam and water circuits against accidental over-pressurization due to fluctuating operating condition or operational faults.

A typical HRSG will have safety valves installed on the:
- LP drum
- IP drum
- HP drum
- LP economizer bypass
- IP steam outlet
- HP steam outlet
- reheater inlet
- reheater outlet

The valves are generally of standard spring-operated design (Fig. 6.6 in Chapter 6 shows typical design and components) and are sized to have a pressure-relieving capacity at least equal to the full load rated steam flow of the boiler. The valves on the IP and HP outlets are often of the power operated types, termed Electromatic Relief Valves (ERVs), which operate by hydraulic or electric power. These valves do not operate as a result of minor excursions in pressure and can be isolated, hence they are only fitted in duplicate with standard spring-operated types.

HRSG level control

A level control system is required to balance the flow of feedwater to the HRSG with the demand for steam flow to the turbines. This requirement is particularly onerous during periods of load change as poor level control can result in water being carried over with the steam into the turbines, causing damage. Conversely, a lack of water to the HRSG circuits can cause overheating and failure of the tube banks. Most HRSGs use a standard three-element analog control system (see Fig. 7.10) measuring:

- drum level
- feedwater flow
- steam flow

These three variables are transmitted as feedback signals to the distributed control system (DCS). One of the prime functions of the control system is to minimize the confusing effects of 'shrink and swell effects'. These are transient effects caused during load changes (Fig. 7.11).

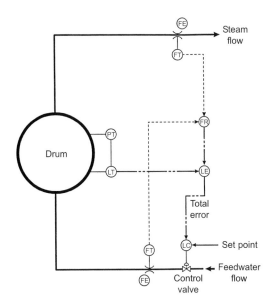

Fig. 7.10 Boiler principles – three element drum level control

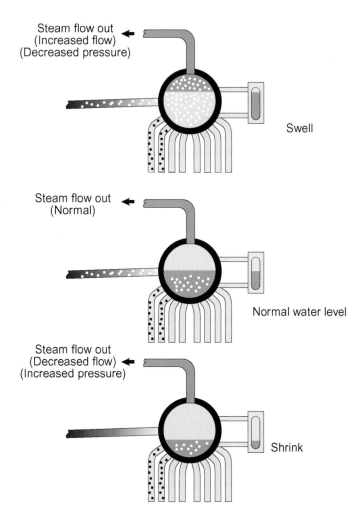

Fig. 7.11 Boiler principles – drum swell and shrink effect

- *Shrink* is the falling of steam drum level during conditions of reduced steam flow and, therefore, increased drum pressure. It is caused by water in the steam drum falling below its saturation temperature. This results in steam bubbles contracting in size, or even condensing back into the water, thereby *lowering* the drum level artificially.
- *Swell* is the opposite effect, experienced during conditions of increased steam demand. The increased demand causes drum pressure to reduce, thereby placing some of the water in the drum at above its saturation temperature for this new reduced pressure. This water will 'flash off' to steam causing the drum level to *rise* artificially.

Steam temperature control

The function of the steam temperature control system is to maintain design superheat temperatures of the steam outlet over the various operating conditions of the HRSG. The typical HRSG type, shown in Fig. 7.6, has two separate control systems: one for control of the superheat steam temperature and the other for control of the reheat outlet superheat temperature. Most modern designs use water-spray desuperheaters (also called *attemporators*), which cool the superheated steam by flashing the spray water (from the HRSG feed system) into steam (see Fig. 7.8).

7.2.3 HRSG terms and definitions

Listed below is some commonly used HRSG terminology.

Baffle	A plate in an HRSG that directs the flow of water, gas, and steam.
Blowdown	Removal of liquids and solids from a vessel or line by the use of a special high-pressure drop valve.
Boiler	An enclosed vessel (as in an HRSG) within which water is heated and circulated to produce steam.
Cascade blowdown	A continuous blowdown system in a boiler that controls water chemistry in all of the HRSG boiler drums by blowing water from a higher-pressure drum to a lower-pressure drum.
Centrifugal separator	A device used in an HRSG steam drum to separate water from steam.
Coils	The pipe system within an HRSG containing the headers and tubes required to exchange heat between exhaust gas and water entering the boilers.

Continuous blowdown	The process of allowing a continuous flow of water to escape from a boiler drum for the purpose of controlling the concentration of treatment chemicals and impurities in the water remaining in the boiler drum.
Deaeration	The removal of non-condensable gas or air from water.
Denox	An abbreviation for the process of reducing oxides of nitrogen gas concentrations (as in the GT (Gas Turbine) exhaust gases).
Diffuser	The part of the GT that straightens and distributes exhaust gas flow uniformly to the HRSG.
Downcomer	A tube or pipe in a boiler system within which water flows downward.
Duct burner	A device located between the GT and the coils that is capable of burning fuel to increase the temperature of exhaust entering the coils.
Economizer	A heat transfer coil in an HRSG that uses the exhaust gas to heat feedwater.
Evaporator	A device, such as a coil, in which water is changed to a vapour by the addition of heat.
Feedwater	Water supplied to a steam-generating unit such as an HRSG.
Fired HRSG	An HRSG with internal fuel gas duct burners capable of producing heat in addition to the heat produced by the GT.
Header	A horizontal pipe that serves as a central connection for two or more tubes.
Intermittent blowdown	The process of periodically allowing water to escape through HRSG drums, valves, and lines to remove sludge and to lower the main steam drum level during start-up, usually applied to the boiler's lower drum.
Riser	A tube in a boiler system within which water flows upward.
Sparge	A system to introduce steam into water through a length of pipe with holes at spaced intervals along the length.
Suction	A pipe or tube feeding into the inlet of a pump.
Superheating	The process of heating steam above its saturation temperature.
Unfired HRSG	An HRSG without internal fuel gas duct burners.

Note: for HRSG Performance Test Code (ANSI/ASME PTC 4.4) go to: http://gatorpwr.che.ufl.edu/cogen/equipment/codes/PTC4.4/Default.asp

7.2.4 HRSG materials

Table 7.3 shows typical HRSG materials.

Table 7.3 Typical HRSG materials

ASME code alloy specification	Nominal composition	Approx. oxidation limit
SA–213	Carbon steel	454 °C
SA–213 T–1	Carbon – 1/2 Mo	482 °C
SA–213 T–11	1–1/4 Cr –1/2 Mo	552 °C
SA–213 T–22	2–1/4 Cr 0 – 1 Mo	579 °C
SA–213 T–9	9 Cr – Mo	626 °C
SA–213 TP 304 H	18 Cr – 8 Ni	704 °C
SA–213 TP 347 H	18 Cr – 10 Ni	704 °C

CHAPTER 8
Materials of Construction

Material properties are of great importance in all aspects of pressure equipment design and manufacture. It is essential to check the up-to-date version of the relevant British Standards, or equivalent, when choosing or assessing a material. The most common steels used for pressure equipment are divided into the generic categories of carbon, alloy, stainless steel, and non-ferrous.

8.1 Plain carbon steels – basic data

Typical properties of plain carbon steels are shown in Table 8.1

Table 8.1 Plain carbon steel: properties

Type	%C	%Mn	Yield, R_e (MN/m²)	UTS, R_m (MN/m²)
Low C steel	0.1	0.35	220	320
General structural steel	0.2	1.4	350	515
Steel castings	0.3	–	270	490

8.2 Alloy steels

Alloy steels have various amounts of Ni, Cr, Mn, or Mo added to improve properties. Typical properties are shown in Table 8.2.

Table 8.2 Alloy steels: properties

Type	%C	Others (%)	R_e (MN/m²)	R_m (MN/m²)
Ni/Mn steel	0.4	0.85 Mn 1.00 Ni	480	680
Ni/Cr Steel	0.3	0.5 Mn 2.8 Ni 1.0 Cr	800	910
Ni/Cr/Mo steel	0.4	0.5 Mn 1.5 Ni 1.1 Cr 0.3 Mo	950	1050

8.3 Stainless steels – basic data

Stainless steel is a generic term used to describe a family of steel alloys containing more than about 11 per cent chromium. The family consists of four main classes, subdivided into approximately 100 grades and variants. The main classes are austenitic and duplex; the other two classes, ferritic and martensitic, tend to have more specialized application and so are not as commonly found in general pressure equipment use. The basic characteristics of each class are given below.

- *Austenitic* The most commonly used basic grades of stainless steel are usually austenitic. They have 17–25% Cr, combined with 8–20% Ni, Mn, and other trace alloying elements that encourage the formation of austenite. They have low carbon content, which makes them weldable. They have the highest general corrosion resistance of the family of stainless steels.
- *Ferritic* Ferritic stainless steels have high chromium content (>17% Cr) coupled with medium carbon, which gives them good corrosion resistance properties rather than high strength. They normally have some Mo and Si, which encourage the ferrite to form. They are generally non-hardenable.
- *Martensitic* This is a high-carbon (up to 2% C), low-chromium (12% Cr) variant. The high carbon content can make it difficult to weld.
- *Duplex* Duplex stainless steels have a structure containing both austenitic and ferritic phases. They can have a tensile strength of up to twice that of straight austenitic stainless steels, and are alloyed with various trace elements to aid corrosion resistance. In general, they are as weldable as austenitic grades, but have a maximum temperature limit, because of the characteristic of their microstructure.

Table 8.3 gives basic stainless steel data.

Table 8.3 Stainless steels basic data

Stainless steels are commonly referred to by their AISI equivalent classification (where applicable).

AISI	Other classifications	Type[+]	Yield F_{ty} (ksi)	[R_e) MPa]	Ultimate F_{tu} (ksi)	[R_m) MPa]	E(%) 50 mm	HRB	%C	%Cr	% others[+]	Properties
302	ASTM A296 (cast), Wk 1.4300, 18/8, SIS 2331	Austenitic	40	[275.8]	90	[620.6]	55	85	0.15	17–19	8–10 Ni	A general purpose stainless steel.
304	ASTM A296, Wk 1.4301, 18/8/LC, SIS 2333, 304S18	Austenitic	42	[289.6]	84	[579.2]	55	80	0.08	18–20	8–12 Ni	An economy grade.
304L	ASTM A351, Wk 1.4306 18/8/ELC, SIS 2352, 304S14	Austenitic	39	[268.9]	80	[551.6]	55	79	0.03	18–20	8–12 Ni	Low C to avoid intercrystalline corrosion after welding.
316	ASTM A296, Wk 1.4436 18/8/Mo, SIS 2243, 316S18	Austenitic	42	[289.6]	84	[579.2]	50	79	0.08	16–18	10–14 Ni	Addition of Mo increases corrosion resistance.
316L	ASTM A351, Wk 1.4435, 18/8/Mo/ELC, 316S14, SIS 2353	Austenitic	42	[289.6]	81	[558.5]	50	79	0.03	16–18	10–14 Ni	Low C weldable variant of 316.

Table 8.3 Cont.

321	ASTM A240, Wk 1.4541, 18/8/Ti, SIS 2337, 321S18	Austenitic	35	[241.3]	90	[620.6]	45	80	0.08	17–19		9–12 Ni	Variation of 304 with Ti added to improve temperature resistance.
405	ASTM A240/A276/A351, UNS 40500	Ferritic	40	[275.8]	70	[482.7]	30	81	0.08	11.5–14.5	1 Mn		A general-purpose ferritic stainless steel.
430	ASTM A240/A276, UNS 43000, Wk 1.4016	Ferritic	50	[344.7]	75	[517.1]	30	83	0.12	14–18	1 Mn		Non-hardening grade with good acid-resistance.
403	UNS S40300, ASTM A176/A276	Martensitic	40	[275.8]	75	[517.1]	35	82	0.15	11.5–13	0.5 Si		Turbine grade of stainless steel.
410	UNS S40300, ASTM A176/A240, Wk 1.4006	Martensitic	40	[275.8]	75	[517.1]	35	82	0.15	11.5–13.5		4.5–6.5 Ni	Used for machine parts, pump shafts, etc.
—	255 (Ferralium)	Duplex	94	[648.1]	115	[793]	25	280 HV	0.04	24–27		4.5–6.5 Ni	Better resistance to SCC than 316.
—	Avesta SAF 2507§, UNS S32750	'Super' duplex 40% ferrite	99	[682.6]	116	[799.8]	~25	300 HV	0.02	25	7Ni, 4Mo, 0.3N		High strength. Max 575 °F (301 °C) due to embrittlement.

* Main constituents only shown.
† All austenitic grades are non-magnetic, ferritic and martensitic grades are magnetic.
§ Avesta trade mark.

8.4 Non-ferrous alloys – basic data

The term 'non-ferrous alloys' is used for those alloy materials that do not have iron as their base element. The main ones used for mechanical engineering applications, with their ultimate tensile strength ranges, are:

- nickel alloys   400–1200 MN/m^2
- zinc alloys   200–360 MN/m^2
- copper alloys   200–1100 MN/m^2
- aluminium alloys   100–500 MN/m^2
- magnesium alloys   150–340 MN/m^2
- titanium alloys   400–1500 MN/m^2

The main non-ferrous alloy materials in use are nickel alloys, in which nickel is frequently alloyed with copper or chromium and iron to produce material with high temperature and corrosion resistance. Typical types and properties are shown in Table 8.4.

Table 8.4 Nickel alloys: properties

Alloy type	Designation	Constituents (%)	UTS (MN/m^2)
Ni–Cu	UNS N04400 ('Monel')	66Ni, 31Cu, 1Fe, 1Mn	415
Ni–Fe	'Ni Io 36'	36Ni, 64Fe	490
Ni–Cr	'Inconel 600'	76Ni, 15Cr, 8Fe	600
Ni–Cr	'Inconel 625'	61Ni, 21Cr, 2Fe, 9Mo, 3Nb	800
Ni–Cr	'Hastelloy C276'	57Ni, 15Cr, 6Fe, 1Co, 16Mo, 4W	750
Ni–Cr (age hardenable)	'Nimonic 80A'	76Ni 20Cr	800–1200
Ni–Cr (age hardenable)	'Inco Waspalloy'	58Ni, 19Cr, 13Co, 4Mo, 3Ti, 1Al	800–1000

8.5 Material traceability

The issue of 'material traceability' is an important aspect of the manufacture of pressure equipment. Most technical codes and standards make provision for quality assurance activities designed to ensure that materials of construction used in the pressure envelope are traceable.

Figure 8.1 shows the 'chain of traceability' that operates for pressure equipment materials. Note that, although all the activities shown are available for use, (i.e. to be specified and then implemented) this does not represent a unique system of traceability suitable for all materials. In practice there are several 'levels' in use, depending both on the type of

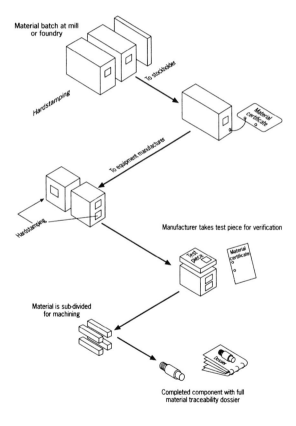

Fig. 8.1 Materials – the chain of traceability

material and the nature of its final application. The most common document referenced in the material sections of pressure equipment specifications is the European Standard EN 10 204: Metallic products – types of inspection documents. It provides for two main 'levels' of certification: Class '3' and Class '2' (see Table 8.5).

Table 8.5 Material traceability: EN10 204 classes

EN 10 204 Certificate type	Document validation by	Compliance with:		Test results included	Test basis	
		The order	'Technical rules'*		Specific	Non-specific
3.1A	I	•	•	Yes	•	
3.1B	M(Q)	•	•	Yes	•	–
3.1C	P	•		Yes	•	
3.2	P + M(Q)	•		Yes	•	–
2.3	M			Yes	•	–
2.2	M			Yes	–	•
2.1	M	•	–	No	–	•

I An independent (third party) inspection organization.

P The purchaser.

M(Q) An 'independent' (normally QA) part of the material manufacturer's organization.

M An involved part of the material manufacturer's organization.

* Normally the 'technical rules' on material properties given in the relevant material standard (and any applicable pressure vessel code).

EN 10 204 and the PED

Although there is no formal link between EN 10 204 and the PED, the PED requires that most pressure equipment materials be provided with a certificate of 'specific product control'. Practically, this coincides with EN 10 204 class 3.1 certificates.

8.6 Materials standards – references*

BS 1503: 1989	*Specification for steel forgings for pressure purposes.* A related standard is ISO 2604/1.
ASTM A273/A273M: 1994	*Specification for alloy steel forgings for high-strength, pressure component application.*
BS 1504: 1984	*Specification for steel castings for pressure purposes.*
ASTM A487/A487M: 1993	*Steel castings suitable for pressure service.*
ASTM A703/A703M: 1994	*Steel castings: general requirements for pressure containing parts.*
BS EN 10130: 1991	*Specification for cold-rolled, low carbon steel flat products for cold forming: technical delivery conditions.*
BS 1501: Part 3: 1990	*Specification for corrosion and heat-resisting steels, plate, sheet, and strip.*
BS 5996: 1993	*Specification for acceptance levels for internal imperfections in steel plate, strip and wide flats, based on ultrasonic testing.* A related standard is EN 160.
BS 3059: Part 1: 1993	*Specification for low tensile carbon steel tubes without specified elevated temperature properties.*
BS 3059: Part 2: 1990	*Specification for carbon, alloy, and austenitic stainless steel tubes with specified elevated temperature properties.* Related standards are ISO 1129, ISO 2604/2, and ISO 2604/3.
DIN 17 245: 1987	*Ferritic steel castings with elevated temperature properties: technical delivery conditions.*
ASTM A356/A356M: 1992	*Specification for steel castings, (carbon low alloy and stainless steel, heavy wall) for steam turbines.*
DIN 17 155: 1989	*Steel plates and strips for pressure purposes.*
BS 3604 Part 1: 1990	*Specification for seamless and electric resistance welded tubes.*
BS 3605: 1992	*Austenitic stainless steel pipes and tubes for pressure purposes.*
ASTM A487/A487M	*Specification for steel castings suitable for pressure service.*
ASTM A430/A430M: 1991	*Austenitic steel forged and bored pipe for high-temperature service.*

* *Note* Many British Standards are in the process of being replaced by European harmonized standards and so may have been recently withdrawn. Currently, however, many of the superseded British Standards are still in regular use in the pressure equipment industry. Appendix 3 of this databook shows the recent status of relevant Directives and harmonized standards. The CEN TC/xxx references may be used to find the current state of development. In all cases it is essential to use the current issue of published technical standards.

CHAPTER 9
Welding and NDT

9.1 Weld types and symbols

Figure 9.1 shows the main types of weld orientation used in pressure equipment construction. Figure 9.2 shows weld preparation terminology. Figures 9.3 and 9.4 show some common British Standard and American welding symbols used.

9.2 Weld processes

Shielded metal arc (SMAW) or manual metal arc (MMA)

This is the most commonly used technique. There is a wide choice of electrodes, metal and fluxes, allowing application to different welding conditions. The gas shield is evolved from the flux, preventing oxidation of the molten metal pool (Fig. 9.5).

Metal inert gas (MIG)

Electrode metal is fused directly into the molten pool. The electrode is, therefore, consumed rapidly, being fed from a motorized reel down the centre of the welding torch (Fig. 9.6).

Tungsten inert gas (TIG)

This uses a similar inert gas shield to MIG, but the tungsten electrode is not consumed. Filler metal is provided from a separate rod fed automatically into the molten pool (Fig. 9.7).

Submerged arc welding (SAW)

Instead of using shielding gas, the arc and weld zone are completely submerged under a blanket of granulated flux (Fig. 9.8). A continuous wire electrode is fed into the weld. This is a common process for welding

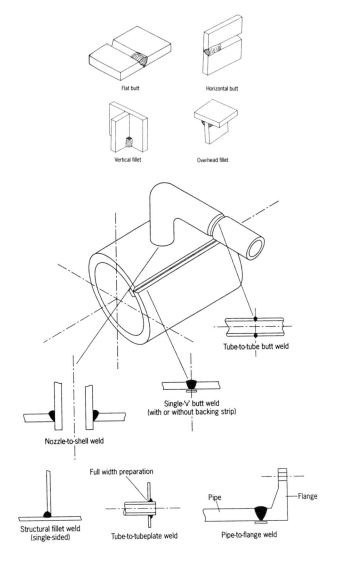

Fig. 9.1 Weld types and orientation

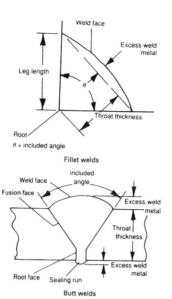

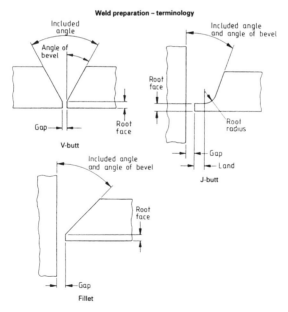

Fig. 9.2 Weld preparations – terminology

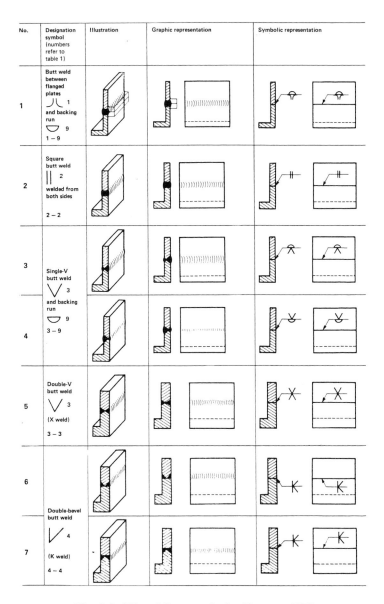

Fig. 9.3 BS welding symbols. (Source: BSI)

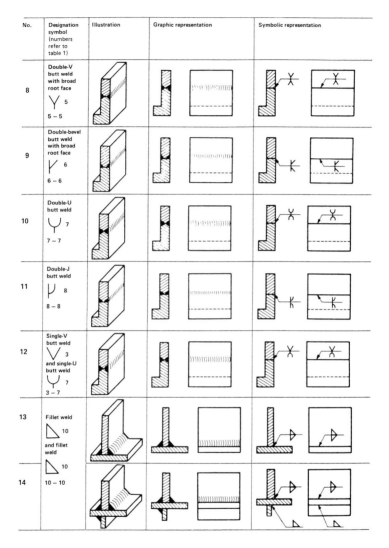

Fig. 9.3 BS welding symbols (cont.)

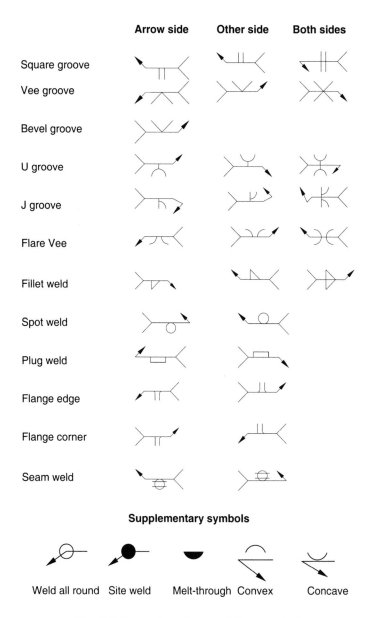

Fig. 9.4 Some American welding symbols

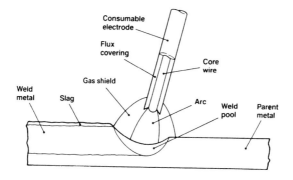

Fig. 9.5 MMA welding

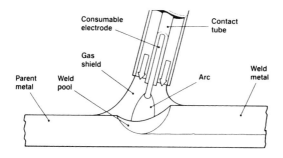

Fig. 9.6 MIG welding

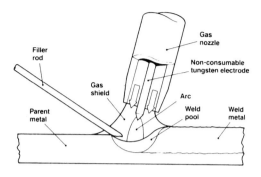

Fig. 9.7 TIG welding

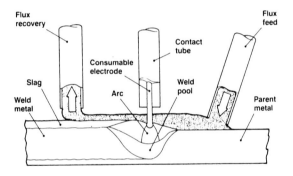

Fig. 9.8 SMA welding (SMAW)

structural carbon or carbon–manganese steel vessels. It is usually automatic with the welding head being mounted on a traversing machine. Long continuous welds are possible with this technique.

Flux-cored arc welding (FCAW)

This is similar to the MIG process, but uses a continuous hollow electrode filled with flux, which produces the shielding gas (Fig. 9.9). The advantage of the technique is that it can be used for outdoor welding, as the gas shield is less susceptible to draughts.

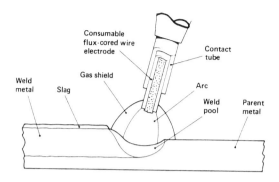

Fig. 9.9 FCA welding (FCAW)

Electrogas welding (EGW)

This is a mechanized electrical process using an electric arc generated between a solid electrode and the workpiece. It has similarities to the MIG process.

Plasma welding (PW)

Plasma welding is similar to the TIG process (Fig. 9.10) A needle-like plasma arc is formed through an orifice and fuses the base metal. Shielding gas is used. Plasma welding is most suited to high-quality and precision welding applications.

9.3 Welding standards and procedures

The standards BS EN 288 and BS EN 287 are used extensively in pressure vessel manufacturing practice. Figure 9.11 shows a diagrammatic representation of how they act as 'control' for the welding activities.

BS EN 288: 1992 *Specification and approval of welding procedures for metallic materials* is divided into six parts and covers the subject of weld procedure specifications (WPSs) and weld procedure qualifications (PQRs or 'tests'). BS EN 288-3 specifies how a WPS is approved by a weld procedure test, to produce an 'approved' or 'qualified' weld procedure. The real core of this standard is to enable the range of approval of a particular PQR to be determined, using the set of tables in the standard.

BS EN 287: 1992 *Approval testing of welders for fusion welding*, relates to the approval testing of the welders. The approval system is based on a set of uniform test pieces and welding positions. As with BS EN 288, there is a

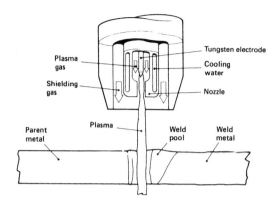

Fig. 9.10 Plasma welding (PW)

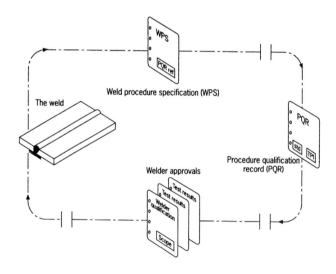

Fig. 9.11 Welding: controlling documentation

range of approval involved. The principle is that, by demonstrating competence on a particular type of weld joint, a welder becomes *qualified* for that weld type and all those weld joints that are 'easier' than the test joint. The standard provides a series of tables to determine where the limits of the ranges of approval lie. BS EN 26520 (ISO 6520): 1992 *Classification of imperfections in metallic fusion welds* is used to classify the imperfections (defects), while BS EN 25817 (ISO 5817): 1992 *Arc-welded joints in steel – guidance on quality levels for imperfections* provides guidance on *quality* levels of these imperfections.

9.4 Destructive testing of welds

9.4.1 Test plates

In contrast to NDT methods, which are essentially predictive techniques, the only *real* way to verify the mechanical integrity of a weld is to test a piece of it to destruction. This may be performed for the purposes of the welding or for the testing of a 'test plate' – properly called a 'production control test plate'. The test plate often represents a pressure vessel longitudinal seam weld. It is tack-welded to the end of the seam preparation, welded at the same time, and then cut off and tested. Figure 9.12 shows the general arrangement.

Test plates are generally specified for vessels or fabricated structures that use very thick material sections, complex alloys or unusual (and fundamentally unproven) weld procedures, or where the essential variables are particularly difficult to control. Although the PED does not mention test plates directly, technical standards (particularly the engineering harmonized standards) do.

9.4.2 The tests

The tests are based on tensile, impact and hardness tests, supplemented by bend tests. The five main tests are shown in Fig. 9.13.

Transverse tensile test

This is a straightforward tensile test *across* the weld. The following points are important:
- *Location of breakage* A good weld should have the same structure and physical properties across the weldment, Heat Affected Zone (HAZ), and parent metal so the failure point could, in theory, be anywhere within the gauge length of the specimen. If the breakage occurs on clearly-defined planes *along the edge* of the weldment/HAZ interface, this could indicate problems of poor fusion or incorrect heat treatment.

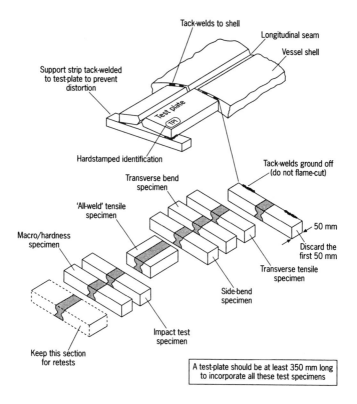

Fig. 9.12 Test plate arrangement – vessel longitudinal seam

- *Tensile strength* There is no real reason why the yield/UTS values should be any different (less) than those of an unwelded specimen of parent metal, so the parent material specification is often used as the acceptance criteria. Some engineers allow an extra tolerance of say 10 per cent on the values (perhaps to allow for inherent uncertainties in the welding process) but there are few current definitive rules on this. The same applies to elongation and reduction of area.
- *Fracture surface appearance* The normal result is a quite dull 'cup and cone type' fracture surface, similar to that expected on a normal tensile test specimen. Marks on the fracture surface may be evidence of metallic or non-metallic (slag) inclusions.

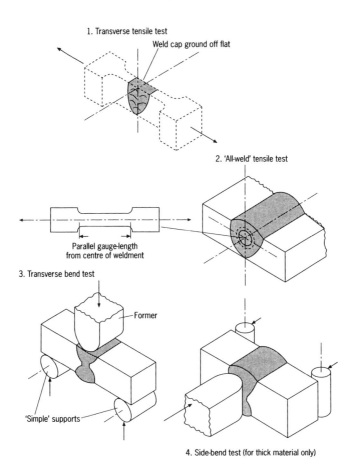

Fig. 9.13 Tensile and bend tests

All-weld tensile test

A tensile test is performed on a specimen machined out longitudinally from the weld. The gauge length diameter is accurately located along the middle of the weld (see Fig. 9.13) so that it is all weldment. The main application of this test is for thick welds that have only single-side access or use a welding technique, which is difficult to control.

Transverse bend test

These are bend tests in which the welded specimen is bent round a roller or former, and are used to (a) verify the soundness of the weldment and HAZ and (b) to check that the HAZ is sufficiently ductile. The transverse test is used for both single and double-sided welds. General acceptance criteria are:
- there should be no visible tearing of the HAZ, or its junction with the weldment;
- cracking or 'crazing' of the material viewed on a macro level ($\times 200$ magnification) is unacceptable – it indicates weaknesses in the metallurgical structure caused by the welding;
- check the *inside* radius of the bend, i.e. the area that is under compressive stress. This may show rippling or 'kinking' around the bend areas, but look for any radial cracks (perpendicular, or at an acute angle to the specimen edge). These are initiated on slip planes in the metal structure and are early indicators of microstructural weakness.

Side bend test

This test checks the soundness of the specimen *across* the weld using a similar former/roller set-up to the transverse bend test. It is generally only carried out on thicker materials (>10 mm). It is particularly useful for manual welds that are more likely to vary along the weld run. General guidelines on acceptance criteria are similar to those for the transverse bend test.

Macro and hardness test

This test consists of etching and polishing a machined transverse slice through the weld, then subjecting it to a close visual examination and hardness check. The specimen is polished using silicon carbide paper (down to P400 or P500 grade) on a rotating wheel. The metal surface is then etched using a mixture of nitric acid and alcohol (for ferritic steels), or hydrochloric acid, nitric acid and water (for austenitic steels).

Once polished and etched, a hardness 'gradient' test is carried out (see Fig. 9.14). This consists of testing and recording hardness values at closely-spaced intervals across the weldment, HAZ, and parent metal regions. The objective is to look for regions that are too hard, or areas in which the hardness gradient is steep, i.e. where hardness varies a lot across a small area.

Other destructive tests that are sometimes done are impact (Charpy 'V' notch) intercrystalline corrosion test (for austenitic stainless steel and pipe welds), fillet weld fracture test, and the fracture toughness ($K{I}c$/COD) test.

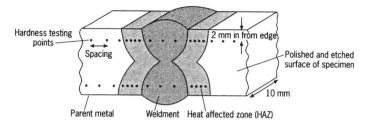

Fig. 9.14 Macro/hardness checks across a weld

9.5 Non-destructive testing (NDT) techniques

NDT techniques are in common use to check the integrity of pressure equipment materials and components. The main applications are plate, forgings, castings, and welds.

9.5.1 Visual examination

Close visual examination can reveal surface cracks and defects of about 0.1 mm and above (see Fig. 9.15). This is larger than the 'critical crack size' for most ferrous materials.

Useful standards

- BS EN 970: 1997: *Non-destructive examination of fusion welds*.
- BS EN 25817: 1992: *Arc-welded joints in steel – guidance on quality levels for imperfections*.
- MSS–SP–55: 1984: *Quality standard for steel casting*.
- BS EN 12062: 1998: *Non-destructive examination of welds – general rules for metallic materials*.

9.5.2 Dye penetrant (DP) testing

This is an enhanced visual technique using three aerosols, a cleaner (clear), penetrant (red), and developer (white). Surface defects appear as a thin red line (see Fig 9.16).

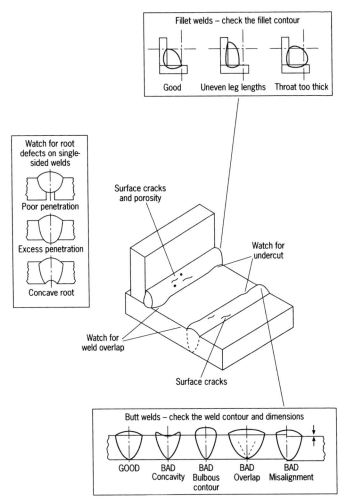

Fig. 9.15 Visual inspection of welds

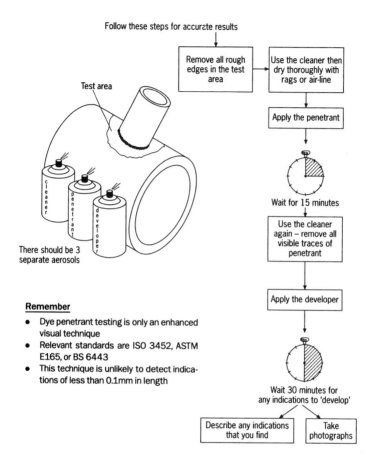

Fig. 9.16 Dye penetrant testing of welds

Useful standards

- ASTM E 165: 1982: *Dye penetrant examination.*
- BS EN 571–1: 1997: *Penetrant testing – general principles.*
- BS EN 1289: 1998: *Penetrant testing of welds – acceptance levels.*
- BS EN ISO 3452–4: 1999: *NDT penetrant testing – equipment.*

9.5.3 Magnetic particle (MP) testing

This works by passing a magnetic flux through the material while spraying the surface with magnetic ink. An air gap in a surface defect forms a discontinuity in the field that attracts the ink, thereby making the crack visible, see Fig. 9.17.

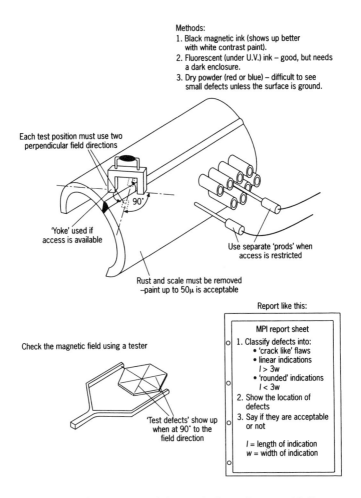

Fig. 9.17 Magnetic particle crack detection – guidelines

Defects are classified into:
- 'crack-like' flaws;
- linear flaws ($l > 3w$);
- rounded flaws ($l < 3w$).

Useful standards

- BS EN 1290: 1998: *Magnetic particle examination of welds*.
- BS EN 1291: 1998: *Magnetic particle testing of welds – acceptance levels*.
- ASTM E 1444: 1984: *Magnetic particle detection for ferromagnetic materials*.

9.5.4 Ultrasonic testing (UT)

Different practices are used for plate, forgings, castings, and welds. The most common technique is the 'A-scope pulse-echo' method (see Fig. 9.18).

UT of plate

Pressure equipment standards contain various 'grades' of acceptance criteria. Plate is tested to verify its compliance with a particular grade

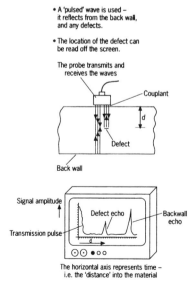

Fig. 9.18 Ultrasonic testing – the 'A-scope' method

specified for the edges and body of the material. Typical criteria are given in Tables 9.1 and 9.2 – these are based on the well-accepted BS 5996. This standard has recently been withdrawn to make room for a harmonized European standard, but its technical content remains in regular use in the pressure equipment industry.

Table 9.1 Typical material 'edge-grades'

Acceptance 'grade'	Single imperfection Max. length (area)	Multiple imperfections, max. no. per 1 m length	Above min. length
E1	50 mm (1000 mm^2)	5	30 mm
E2	30 mm (500 mm^2)	4	20 mm
E3	20 mm (100 mm^2)	3	10 mm

Table 9.2 Typical material 'body-grades'

Acceptance 'grade'	Single imperfection max. area (approximate)	Multiple imperfections, max. no. per 1 m square	Above min. size (area)
B1	10 000 mm^2	5	10 mm x 20 mm (2500 mm^2)
B2	5000 mm^2	5	75 mm x 15 mm (1250 mm^2)
B3	2500 mm^2	5	60 mm x 12 mm (750 mm^2)
B4	1000 mm^2	10	35mm x 8 mm (300 mm^2)

Useful standard
- BS 5996: 1993: *Specification for acceptance levels for internal imperfections in steel plate, based on ultrasonic testing* (equivalent to EN 160).

UT of castings
Casting discontinuities can be either planar or volumetric. Separate gradings are used for these when discovered by UT technique. The areas of a

pressures equipment casting are typically divided into critical and non-critical areas, and by thickness 'zones' (Fig. 9.19). Typical grading criteria are as shown in Tables 9.3 and 9.4.

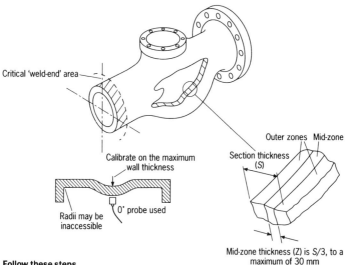

Follow these steps
- Check the suitability of the casting for an ultrasonic technique (the sound permeability)
- Do a visual inspection
- Do a preliminary scan – looking for both types of discontinuity
- Assess the planar discontinuities
- Assess the non-planar discontinuities (find their upper and lower 'bounds' and then delineate their area)
- Classify the results into grades, using the tables shown
- Report accurately

Fig. 9.19 Ultrasonic tests on castings

Table 9.3 Castings: planar discontinuities

Planar discontinuities	Grade			
	1	2	3	4
Max. 'through-wall' discontinuity size	0 mm	5 mm	8 mm	11 mm
Max. area of a discontinuity	0 mm	75 mm^2	200 mm^2	360 mm^2
Max. total area* of discontinuities	0 mm	150 mm^2	400 mm^2	700 mm^2

Table 9.4 Casting: non-planar discontinuities

Non-planar discontinuities	Grade			
	1	2	3	4
Outer zone Max. size	0.2Z	0.2Z	0.2Z	0.2Z
Out zone Max. total area*	250 mm^2	1000 mm^2	2000 mm^2	4000 mm^2
Mid zone Max. size	0.1S	0.1S	0.15S	0.15S
Mid zone Max. total area*	12 500 mm^2	20 000 mm^2	31 000 mm^2	50 000 mm^2

* All discontinuity levels are per unit (10 000 mm^2) area

Useful standards

- ASTM A609: 1991: *Practice for ultrasonic examination of castings*.
- BS 6208: 1990: *Methods for ultrasonic testing of ferrite steel castings, including quality levels.*

UT of welds

Weld UT has to be a well-controlled procedure, because the defects are small and difficult to classify. Ultrasonic scans may be necessary from several different directions, depending on the weld type and orientation.

The general techniques, shown in Figs 9.20 and 9.21, are:

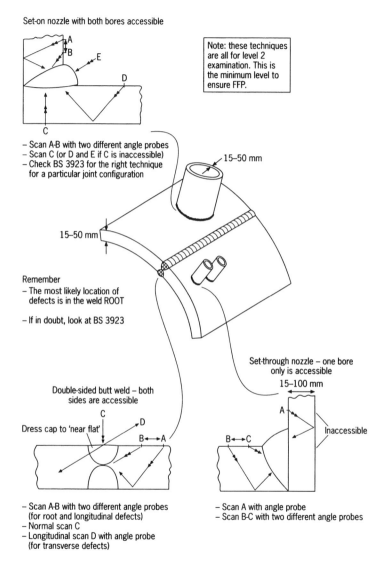

Fig. 9.20 Ultrasonic testing of butt and nozzle welds

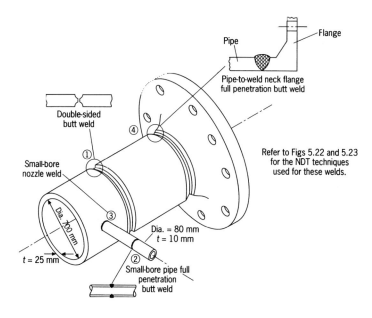

Fig. 9.21 Pipework – welds for volumetric NDT

- surface scan using normal (0°) probe;
- transverse scan (across the weld) to detect *longitudinal* defects;
- longitudinal scan (along the weld direction) to detect *transverse* defects.

Useful standards

- BS EN 583-1: 1999: *Ultrasonic examination – general principles*.
- BS EN 1714: 1998: *Ultrasonic examination of welded joints*.
- BS EN 1712: 1997: *Ultrasonic examination of welded joints – acceptance levels*.
- BS EN 1713: 1998: *Ultrasonic examination – characterisation of indications in welds*.

9.5.5 Radiographic testing (RT)

Radiography is widely used for NDT of pressure equipment components and welds.
- X-rays are effective on steel up to a thickness of approximately 150 mm.
- Gamma (γ) rays can also be used for thickness of 50–150 mm, but definition is not as good as with X-rays.

Techniques

For tubular components a single or double-wall technique may be used. Figures 9.22 and 9.23 show the typical way in which a RT technique is specified for large and small pipe butt welds.

Penetrameters

Penetrameters, or image quality indicators (IQIs), check the sensitivity of a radiographic technique to ensure that any defects present will be visible. The two main types are the 'wire' type and 'hole' type. Figure 9.24 shows the ASTM 'hole-type'.

Useful standards

Radiography techniques
- BS EN 444: 1994: *NDT General principles for RG examination of metallic materials*.
- BS EN 1435: 1997: *Radiographic examination of welded joints*.
- BS 2737: 1995: *Terminology of internal defects in castings as revealed by radiography*.
- DIN 54111: *Guidance for the testing of welds with X-rays and gamma-rays*.
- ISO 4993: *RG examination of steel castings*.
- ISO 5579: *RG examination of metallic materials by X and gamma radiography*.

Standards concerned with image clarity
- BS EN 462: 1994: *Image quality of radiographs*.
- DIN 55110: *Guidance for the evaluation of the quality of X-ray and gamma-ray radiography of metals*.
- ASTM E142: 1992: *Methods for controlling the quality of radiographic testing*.
- BS EN 1330: 1997: *NDT terminology*.

A double-wall, double image technique is used.
The technique looks like this:

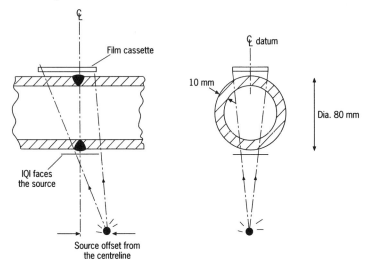

and is specified like this:

Specification	Explanation
Double-wall/image	Two weld 'thicknesses' show
Technique no.13	A reference to BS 2910 which lists techniques 1–16
Class B	Double-wall techniques are inferior to single-wall methods
Fine film	BS 2910 mentions the use of fine or medium grades
Density 3.5–4.5	The 'degree of blackness' of the image

Notes
- The IQI is BS 3971 type I (BS EN 462–1)
- For 2 × 10 mm thickness, the maximum acceptable sensitivity given in BS 2910 is 1.6%. This means that the 0.2 mm wire should be visible
- A minimum of three films are required, spaced at 0°, 120°, and 240° from the datum
- The maximum OD for this technique is 90 mm
- X-ray gives better results than gamma in techniques like this
- The source is offset to prevent superimposed images

Fig. 9.22 RT technique for small bore butt weld

A single-wall X-ray technique is used:
- the large bore allows access
- the source-outside technique is commonly used for large pipe

The technique looks like this:

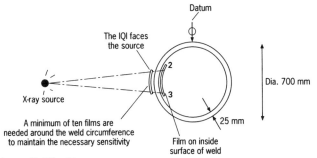

and is specified like this:

Specification	Explanation
Single wall	Only one weld 'thickness' shows on the film
Technique no.1	A reference to BS 2910 which lists techniques nos 1–16
Class A	X-ray and single wall techniques give the best (class A) results
Fine film	BS 2910 mentions the use of fine or medium film grades
X-220 kV	The X-ray voltage depends on the weld thickness

Notes
- The IQI is BS 3971 type I (BS EN 462–1)
- For 25 mm weld thickness the maximum acceptable sensitivity given in BS 2910 is 1.3%. This means that the 0.32 mm wire should be visible
- The IQI should be located in the area of worst expected sensitivity

Film marking (using lead symbols)

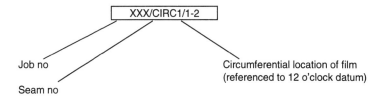

Fig. 9.23 RT technique for large bore butt weld

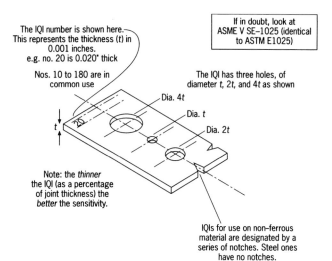

IQI designation	Sensitivity	Visible hole*
1–2t	1	2t
2–1t	1.4	1t
2–2t	2.0	2t
2–4t	2.8	2t
4–2t	4.0	2t

* The hole that must be visible in order to ensure the sensitivity level shown

Fig. 9.24 The ASTM hole-type penetrameter

9.6 NDT acronyms

Pressure equipment examination procedures, reports, and general literature are full of acronyms. The most common ones are listed below.

AE	Acoustic Emission
AFD	Automated Flaw Detection
A-Scan	Amplitude Scan
ASNT	American Society for Non-destructive Testing
ASTM	American Society for Testing and Materials
B-Scan	Brightness Scan
BVID	Barely Visible Impact Damage
CDI	Crack Detection Index
CRT	Cathode Ray Tube
C-Scan	Contrast Scan
CSI	Compton Scatter Imaging
CTM	Coating Thickness Measurement
CW	Continuous Wave/Compression Wave
DAC	Distance Amplitude Correction
dB	Decibel
DGS	Distance, Gain, Size (Diagram)
DPEC	Deep Penetration Eddy Currents
EC	Eddy Current
ECII	Eddy Current Impedance Imaging
EPS	Equivalent Penetrameter Sensitivity
ET (ECT)	Eddy Current Testing
FFD	Focus-to-Film Distance
FSH	Full-Scale Height
HAZ	Heat Affected Zone
HDR	High-Definition Radiography
HVT	Half Value Thickness
IF	Industrial Fibrescope
IQI	Image Quality Indicator
IV	Industrial Video-imagescope
LD	Linear Detectors
LFECA	Low-Frequency Eddy Current Array
LPI	Liquid Penetrant Inspection
LW	Longitudinal Wave
MFL	Magnetic Flux Leakage
MPI	Magnetic Particle Inspection
MPT	Magnetic Particle Testing

MR	Microradiography
MRI	Magnetic Resonance Imaging
MT	Magnetic Testing
NDA	Non-Destructive Assessment
NDE	Non-Destructive Examination
NDI	Non-Destructive Inspection
NDT	Non-Destructive Testing
NMR	Nuclear Magnetic Resonance
PA	Peak Amplitude
PDRAM	Pulsed Digital Reflection Acoustic Microscopy
POD	Probability Of Detection
P-Scan	Projection Scan
PT	Penetrant Testing
PVT	Pulse Video Thermography
QNDE	Quantitative Non-Destructive Evaluation
RFET	Remote Field Eddy Current Testing
ROI	Region Of Interest
ROV	Remotely-Operated Vehicle
RT	Radiographic Testing
RT	Real Time
RTUIS	Real Time Ultrasonic Imaging System
RVI	Remote Visual Inspection
RVT	Remote Visual Testing
SAM	Scanning Acoustic Microscopy
SDT	Static Deflection Techniques
SEM	Scanning Electron Microscopy
SFD	Source-to-Film Distance
SH	Horizontally Polarized Shear Waves
SI	Sensitivity Indicator
SIT	Simulated Infrared Thermography
SMNR	Signal-to-Material Noise Ratio
SNR	Signal-to-Noise Ratio
SPATE	Stress Pattern Analysis by Thermal Emission
TDR	Time-Domain Reflectometry
TOFD	Time-Of-Flight Diffraction
TSE	Total Spectral Energy
TW	Transverse Wave
US	Ultrasonic
UT	Ultrasonic Testing
VAP	Variable Angle (Ultrasonic) Probe
VT	Visual Testing

Welding and NDT 225

WFMPI	Wet Fluorescent Magnetic Particle Inspection
WIR	Work and Inspection Robot
WT	Wall Thickness

9.7 NDT: vessel code applications

Pressure vessel (and other pressure equipment) codes vary in the way that they address both the extent of NDT required and the acceptance criteria that are applied to the results. Tables 9.5 and 9.6 summarize NDT requirements given in the vessel codes covered in chapter 4. Note that these are indicative requirements only: code inquiry cases may exist that can clarify (or even amend) these requirements.

Table 9.5 NDT requirements: pressure vessel codes (quick reference)

PD 5500	***Cat 1 vessels*** • 100% RT or UT on type A joints (mandatory). • 100% RT or UT on type B joints above a minimum thickness (see Table 5.6.4.1.1). • 100% crack detection (DP or MPI) on all type B joints and attachment welds. • Crack detection on type A joints is optional (by agreement only). ***Cat 2 vessels*** • Minimum 10% RT or UT of the aggregate length of type A longitudinal and circumferential seams to include all intersections and areas of seam on or near nozzle openings. • Full RT or UT on the nozzle weld of one nozzle from every ten. • 100% crack detection (DP or MPI) on all nozzle welds and compensation plate welds. • 10% crack detection (DP or MPI) on all other attachment welds to pressure parts. ***Cat 3 vessels*** • Visual examination only: no mandatory NDT. • Weld root grind-back must be witnessed by the Inspecting Authority.
ASME VIII	***Div 1 vessels*** The extent of NDT is not straightforward. Look first at part UW-11 then follow the cross-references. Generally, the following welds require 100% RT: • Butt welds in material ≥ 38 mm thick. • Butt welds in unfired boiler vessels operating at ≥ 3.4 bar g.

Table 9.5 Cont.

	Other welds are typically subject to spot RT prescribed by UW–52. ***Div 2 vessels*** • All pressure shell welds require 100% RT.
TRD	• Longitudinal welds (LN): Generally 100% RT or UT is required. • Circumferential welds (RN): Between 10% and 100% RT or UT depending on material and thickness. • T-Intersections (St) in butt welds: Generally 100% RT or UT. • Nozzle and fillet welds: Generally 100% DP or MPI. The complete details of NDT extent are given in AD-Merkblatt HP O and HP 5/3 tables 1, 2, 3.

Table 9.6 Pressure vessel codes: defect acceptance criteria (quick reference)

PD 5500 (Cat 1)

Visual examination		*Radiographic examination*	
Planar defects: Weld fit-up 'gap': Undercut: Throat thickness error: Excess penetration: Weld 'sag':	Not permitted max 2 mm max 1 mm +5/-1 mm max 3 mm max 1.5 mm	Cracks/lamellar tears Lack of fusion, any type Lack of root penetration	Not permitted
		Porosity: max size of isolated pores is 25% of material thickness, up to a maximum diameter, which depends on the material thickness.	
Root concavity:	max 1.5 mm	Wormholes: max 6 mm length × 1.5 mm width.	
Weld cap overlap:	Not permitted	Inclusions: allowable dimensions of solid inclusions depend on the position of the inclusion and the type of weld [refer to PD5500 Table 5.7(1)].	

Refer to Tables 5.7(1) and 5.7(3) of PD 5500 for full details.
Ultrasonic defect acceptance criteria are given in Table 5.7(2)

Table 9.6 Cont.

ASME VIII	
Radiographic examination (100%)	**'Spot' radiographic examination**
Cracks Lack of fusion, any type } Not permitted Lack of penetration	Cracks Lack of fusion } Not permitted Lack of penetration
Rounded indications: see ASME VIII appendix 4	Slag or cavities: max length 2/3 t to maximum of 19 mm
'Elongated' indications (e.g. slag or inclusions): max length is 6 mm to 18 mm, depending on the material thickness	Porosity: no acceptance criteria
See UW–51 for full details	See UW–52 for full details
TRD	
Radiographic examination	**Ultrasonic examination**
Cracks Lack of side wall fusion } Not permitted	Longitudinal flaws: allowable up to a length approximately equal to material thickness. See HP 5/3 Table 5
Incomplete root fusion: not permitted on single-sided welds	Transverse flaws: max of three per metre length. See HP 5/3 Clause 4.4
Solid and gaseous 'inclusions': max length 7 mm (t< 10 mm) or 2/3 t for 10 mm < t< 75 mm	
Tungsten inclusions: max length 3–5 mm depending on metal thickness	
See AD - Merkblatt HP 5/3 for full details	

9.8 NDT standards* and references

- BS 709: 1983: *Methods of destructive testing fusion welded joints and weld metal in steel.*
- BS 2633: 1987: *Specification for Class I arc welding of ferritic steel pipework for carrying fluids.*
- BS 2971: 1991: *Specification for Class II arc welding of carbon steel pipework for carrying fluids.*
- BS 4570: 1985: *Specification for fusion welding of steel castings.*
- BS EN 288 Parts 1–8: 1992: *Specification and approval of welding procedures for metallic materials.*
- BS EN 287-1: 1992: *Approval testing of welders for fusion welding.*

- BS EN 26520: 1992: *Classification of imperfections in metallic fusion welds, with explanations.* This is an identical standard to ISO 6520.
- BS EN 25817: 1992: *Arc welded joints in steel – guidance on quality levels for imperfections.*
- BS 5289: 1983: Code of practice. *Visual inspection of fusion welded joints.*
- BS 4124: 1991: *Methods for ultrasonic detection of imperfections in steel forgings.*
- BS 1384 Parts 1 and 2. *Photographic density measurements.*
- PD 6493: 1991: *Guidance on methods for assessing the acceptability of flaws in fusion welded structures* (recently withdrawn).

** **Note** Many British Standards are in the process of being replaced by European harmonized standards and so may have been recently withdrawn. Currently, however, many of the superseded British Standards are still in regular use in the pressure equipment industry. Appendix 3 of this databook shows the recent status of relevant Directives and harmonized standards. The CEN TC/xxx references may be used to find the current state of development. In all cases it is essential to use the current issue of published technical standards.*

CHAPTER 10

Failure

10.1 How pressure equipment materials fail

There is no single, universally accepted explanation covering the way that metallic materials fail. Figure 10.1 shows the generally accepted phases of failure. Elastic behaviour, up to yield point, is followed by increasing amounts of irreversible plastic flow. The fracture of the material starts from the point in time at which a crack initiation occurs and continues during the propagation phase until the material breaks.

There are several approaches to both the characteristics of the original material and the way that the material behaves at a crack tip, see Fig. 10.2. Two of the more common approaches that are applicable to pressure

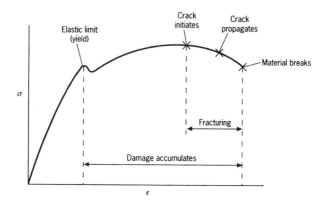

Fig. 10.1 Phases of failure

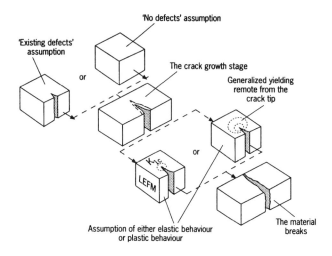

Fig. 10.2 Material failure – the approaches

equipment applications are:
- the linear elastic fracture mechanics (LEFM) approach with its related concept of fracture toughness (K_{1c}) parameter (a material property);
- fully plastic behaviour at the crack tip, i.e. 'plastic collapse' approach.

A useful standard is ASTM E399

10.1.1 LEFM method

This method is based on the 'fast fracture' equation

$$K_{1c} = K_1 \equiv YS\sqrt{(\pi a)}$$

where
K_{1c} = plane strain fracture toughness
K_1 = stress intensity factor
a = crack length
Y = dimensionless factor based on geometry
S = stress level

Typical Y values used are shown in Fig. 10.3.

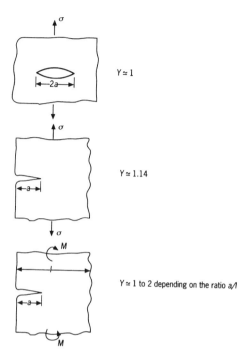

Fig. 10.3 LEFM – 'rule of thumb' Y factors

10.1.2 Multi-axis stress states

When stress is not uniaxial (as in many pressure equipment components), yielding is governed by a combination of various stress components acting together. There are several different 'approaches' as to how this happens.

Von Mises criterion (or 'distortion energy' theory)

This theory states that yielding will take place when

$$\frac{1}{2}^{\frac{1}{2}}\left[(S_1-S_2)^2+(S_2-S_3)^2+(S_3-S_1)^2\right]^{\frac{1}{2}} = \pm S_y$$

where S_1, S_2, S_3 are the principal stresses at a point in a component. This is a useful theory for ductile metals.

Tresca criterion (or maximum shear stress theory)

$$\frac{(S_1-S_2)}{2} \text{ or } \frac{(S_2-S_3)}{2} \text{ or } \frac{(S_3-S_1)}{2} = \pm\frac{S_y}{2}$$

This is also a useful theory for ductile materials.

Maximum principal stress theory

This is a simpler theory which is a useful approximation for brittle metals. The material fails when

S_1 or S_2 or $S_3 = \pm S_y$

10.2 Fatigue

Ductile materials can fail at stresses significantly less than their rated yield strength if they are subject to fatigue loadings. Fatigue data are displayed graphically on a *S–N* curve (see Fig. 10.4). Some pressure equipment materials exhibit a 'fatigue limit', representing the stress at which the material can be subjected to (in theory) an infinite number of cycles without exhibiting any fatigue effects. This fatigue limit is influenced by the size and surface finish of the specimen, as well as the material's properties. Figure 10.5 shows types of stress loading.

Characteristics of fatigue failures are:
- visible crack-arrest and 'beach mark' lines on the fracture face;
- striations (visible under magnification) – these are the result of deformation during individual stress cycles;
- an initiation point such as a crack, defect, or inclusion, normally on the surface of the material.

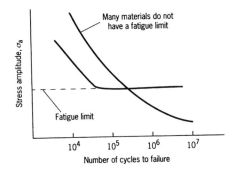

Fig. 10.4 A fatigue 'S–N' curve

Stresses in engineering components are rarely static – they often vary with time (t). The four main classifications are as shown below:

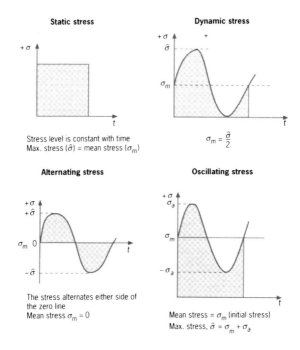

Fig. 10.5 Types of stress loading

10.2.1 *Typical pressure equipment material fatigue limits*

Fatigue limits are a highly controversial topic and various codes and standards take different views. Table 10.1 shows some commonly used 'rules of thumb', in this case in imperial units.

Table 10.1 Typical fatigue limits (rules of thumb only) in imperial units

Material	Tensile strength (S_m) x 1000 lb/in^2	Approx fatigue limit x 1000 lb/in^2
Low carbon steel	50–140	20–70
Cast iron	20–45	6–16
Cast steel	60–75	23–30
Titanium	85–90	40–45
Aluminium	18–35	5–10
Brass	25–75	7–19
Copper	30–45	11–15

10.2.2 Fatigue strength – rules of thumb

The fatigue strength of a material varies significantly with the size and shape of section and the type of fatigue stresses to which it is subjected. Some 'rules of thumb' for pressure equipment materials are shown in Table 10.2. Note how they relate to yield (S_y) and ultimate (S_m) values in pure tension.

Table 10.2 Fatigue strength: (rules of thumb only)

		Steel (structural)	Steel (hardened and tempered)	Cast Iron
Bending	$S_{w(b)}$	$0.5 S_m$	$0.45 S_m$	$0.38 S_m$
	$S_{a(b)}$	$0.75 S_m$	$0.77 S_m$	$0.68 S_m$
	$S_{y(b)}$	$1.5 S_y$	$1.4 S_y$	–
Tension	S_w	$0.45 S_m$	$0.4 S_m$	$0.25 S_m$
	S_a	$0.59 S_m$	$0.69 S_m$	$0.4 S_m$
Torsion	$S_{sw(t)}$	$0.35 S_m$	$0.3 S_m$	$0.35 S_m$
	$S_{sa(t)}$	$0.38 S_m$	$0.5 S_m$	$0.56 S_m$
	$S_{sy(t)}$	$0.7 S_m$	$0.7 S_y$	–

where:

$S_{w(b)}$ = fatigue strength under alternating stress (bending);

$S_{a(b)}$ = fatigue strength under fluctuating stress (bending);

$S_{y(b)}$ = yield point (bending);

S_w = fatigue strength under alternating stress (tension);

S_a = fatigue strength under fluctuating stress (tension);

S_y = yield point (tension);

$S_{sw(t)}$ = fatigue strength under alternating stress (torsion);

$S_{sa(t)}$ = fatigue strength under fluctuating stress (torsion);

$S_{sy(t)}$ = yield point (torsion).

European equivalent (SI) units are shown in Table 10.3. Material strength definitions and equivalent units in use in Europe are as shown.

Table 10.3 Strength units: European equivalents

	Yield strength	*Ultimate tensile strength*	*Modulus*
USCS (US imperial)	F_{ty} (ksi) or S_y (lb/in^2)	F_{tu} (ksi) or S_m (lb/in^2)	E(lb/in^2 × 10^6)
SI/European	R_e (MN/m^2)	R_m (MN/m^2)	E(N/m^2 x 10^9)

Conversions are:
1 ksi = 1000 psi = 6.89 MPa = 6.89 MN/m ^{2} = 6.89 N/mm^2
(see also Fig. 7.4 for pressure conversions)

10.3 Creep

Creep is one of the more complex degradation mechanisms found in pressure equipment. It *is* a specialist subject, but there are some straightforward guidelines.

Basics of creep

What is creep?

Creep is a degradation mechanism found in steels and other pressure equipment materials.

What is it dependent on?

Creep is dependent on two parameters: temperature and time. For steels, creep rarely occurs below about 390 °C and can get progressively worse as the temperature increases above this. Creep damage also increases (more or less linearly) with the time that material is exposed to these high temperatures. Other factors, such as corrosive atmospheres and fatigue conditions, can act to make the situation worse.

What damage does creep cause?

The main effect is a permanent and progressively worsening reduction in tensile strength. The creep mechanism causes the metal structure to 'flow' (on a macro/microscopic scale) leaving holes or *voids* in the material matrix. Structures under pressure stress, therefore, can deform and then fail.

How is creep detected?

Using metallographic replication (often simply termed a 'replica'). This is a non-destructive technique that enables a visual examination of the material's grain structure to look for the voids that are evidence of creep damage. The procedure comprises:

- prepare the metal surface using mechanical cleaning/grinding and a chemical etch;
- bond a special plastic tape to the surface and allow the adhesive to cure;
- remove the tape, transferring a replica of the metal surface to the tape;
- examine the replica under low magnification and compare with reference microstructure pictures.

Note that the results of replica tests are assessed in a qualitative way, comparing the number and size of voids with those on reference pictures for the material under consideration. If a material fails under this test, the next step is to perform full (destructive) tests for tensile strength on a trepanned sample taken from the component.

Where is creep likely in pressure equipment?

Common areas are:

- **Boilers**

    Traditional boilers and HRSGs – Superheater and reheater headers

    High pressure boilers – Platen superheater tube banks

- **Valves**

    Valve bodies and seats on high-temperature systems

- **Turbine casings**

    Stator blade casing diaphragms at the hot end of the casings

What is a typical creep-resistant alloy?

Main steam pipelines for fossil or gas-fired power stations containing superheated steam at 520 °C+ commonly use a DIN 14MoV63 alloy (or a comparable 0.5%Cr, 0.5%Mo, 0.25%V material). This has documented creep-resistant properties up to 100 000 h of exposure.

Glossary of creep testing terms

- **Creep**

  Creep is defined as deformation that occurs over a period of time when a material is subjected to constant stress at constant temperature. In metals, creep usually occurs only at elevated temperatures. Creep at room temperature is more common in plastic materials and is called cold flow or deformation under load. Data obtained in a creep test are usually presented as a plot of creep versus time, with stress and temperature constant. The slope of the curve is creep rate and the end point of the curve is time for rupture. The creep of a material can be divided into three stages. The first stage, or primary creep, starts at a rapid rate and slows with time. The second stage (secondary) creep has a relatively uniform rate. Third stage (tertiary) creep has an accelerating creep rate and terminates by failure of material at time for rupture.

- **Creep limit**

  An alternative term for creep strength.

- **Creep rate**

  Creep rate is defined as the rate of deformation of a material subject to stress at a constant temperature. It is the slope of the creep versus time diagram obtained in a creep test. Units usually are mm/h or % of elongation/h. Minimum creep rate is the slope of the portion of the creep versus time diagram corresponding to secondary creep.

- **Creep recovery**

  Creep recovery is the rate of decrease in deformation that occurs when load is removed after prolonged application in a creep test. Constant temperature is maintained to eliminate effects of thermal expansion, and measurements are taken from zero time load to eliminate elastic effects.

- **Creep rupture strength**

  This is stress required to cause fracture in a creep test within a specified time. An alternative term is 'stress rupture strength'.

- **Creep strength**

  Creep strength is the maximum stress required to cause a specified amount of creep in a specified time. This term is also used to describe maximum stress that can be generated in a material at constant temperature under which creep rate decreases with time. An alternative term is 'creep limit'.

- **Creep test**

    This is a method for determining creep or stress relaxation behaviour. To determine creep properties, material is subjected to prolonged constant tension or compression loading at constant temperature. Deformation is recorded at specified time intervals and a creep versus time diagram is plotted. The slope of the curve at any point is the *creep rate*. If failure occurs, it terminates the test and time for rupture is recorded. If the specimen does not fracture within a test period, creep recovery may be measured. To determine the *stress relaxation* of a material, a specimen is deformed a given amount and the decrease in stress over a prolonged period of exposure at constant temperature is recorded.

10.4 Corrosion

10.4.1 Types of corrosion

There are three basic types of corrosion relevant to generic pressure equipment applications:
- chemical corrosion;
- galvanic corrosion;
- electrolytic corrosion.

To complicate matters there are a variety of sub-types, some hybrids, and some which are relevant mainly to boiler applications.

Chemical corrosion

The cause of this corrosion is attack by chemical compounds in a material's environment. It is sometimes referred to as 'dry' corrosion or oxidation.

Galvanic corrosion

This is caused by two or more dissimilar metals in contact in the presence of a conducting electrolyte. One material becomes anodic to the other and corrodes, see Fig. 10.6.

The tendency of a metal to become anodic or cathodic is governed by its position in the electrochemical series. This is, strictly, only accurate for pure metals rather than metallic compounds and alloys, see ASTM G135 and ASTM G102. A more general guide to the galvanic corrosion attack of common engineering materials is given in Fig. 10.7.

Electrolytic corrosion

This is sometimes referred to as 'wet' or 'electrolytic' corrosion. It is similar to galvanic corrosion in that it involves a potential difference and an

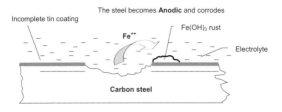

Fig. 10.6 Galvanic corrosion

GALVANIC CORROSION ATTACK – GUIDELINES

Corrosion of the materials in each column is increased by contact with the materials in the row when the corresponding box is shaded.

Material	Steel and Cl	Stainless steel 18% Cr	Stainless steel 11% Cr	Inconel Ni alloys	Cu/Ni and bronzes	Cu and brass	PbSn and soft solder	Silver solder	Mg alloys	Chromium	Titanium	Al alloys	Zinc
Steel and Cl		▓	▓	▓	▓	▓		▓		▓			
Stainless steel 18% Cr								▓					
Stainless steel 11% Cr		▓						■					
Inconel/Ni alloys													
Cu/Ni and bronzes		▓	▓	▓									
Cu and brass		▓	▓	▓	▓								
PbSn and soft solder		▓	▓	▓	▓	▓						▓	
Silver solder		▓	▓	▓	▓	▓							
Mg alloys													
Chromium													
Titanium													
Al alloys		▓	▓	▓	▓	▓	▓	▓					
Zinc												▓	

Example: The corrosion rate of silver solder is increased when it is placed in contact with 11% Cr stainless steel.

Fig. 10.7 Galvanic corrosion attack – guidelines

electrolyte but it does not need to have dissimilar materials. The galvanic action often happens on a microscopic scale. Examples are:
- pitting of castings due to galvanic action between different parts of the crystals (which have different composition);
- corrosion of castings due to grain boundary corrosion (Fig. 10.8).

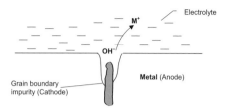

Fig. 10.8 Electrolytic corrosion

Crevice corrosion

This occurs between close-fitting surfaces, crevice faces, or anywhere where a metal is restricted from forming a protective oxide layer (Fig. 10.9) Corrosion normally propagates in the form of pitting. Examples are:
• corrosion in crevices in seal welds;
• corrosion in lap-joints used in fabricated components and vessels.

Stress corrosion

This is caused by a combination of corrosive environment and tensile loading. Cracks in a material's brittle surface layer propagate into the material, resulting in multiple bifurcated (branching) cracks. Examples are:
• failure in stainless steel pipes and bellows in a chlorate-rich environment;
• corrosion of austenitic stainless steel pressure vessels.

Corrosion fatigue

This is a hybrid category in which the effect of a corrosion mechanism is increased by the existence of a fatigue condition. Seawater, fresh water, and even air can reduce the fatigue life of a material.

Intergranular corrosion

This is a form of local anodic attack at the grain boundaries of crystals due to microscopic difference in the metal structure and composition. An example is:
• 'weld decay' in un-stabilized austenitic stainless steels.

Erosion–corrosion

Almost any corrosion mechanism is made worse if the material is subject to simultaneous corrosion and abrasion. Abrasion removes the protective passive film that forms on the surface of many metals, exposing the underlying metal. An example is:
• the walls of pipes containing fast-flowing fluids and suspended solids.

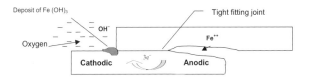

Fig. 10.9 Crevice corrosion

10.4.2 Useful references

Websites

National Association of Corrosion Engineers (NACE), go to:
www.nace.org
For a list of corrosion-related links, go to:
www.nace.org/corlink/corplinkindex.htm

Standards

1 ASTM G15: 1999 *Standard terminology relating to corrosion and corrosion testing.*
2 ASTM G135: 1995 *Standard guide for computerized exchange of corrosion data for metals.*
3 ASTM G119: 1998 *Standard guide for determining synergism between wear and corrosion.*
4 ASTM G102: 1999 *Standard practice for calculation of corrosion rates and related information from electrochemical measurements.*

10.5 Boiler failure modes

Steam boilers can exhibit various types of failure mechanism, caused by corrosion and other factors. Table 10.4 is a 'quick reference' of the main types.

Table 10.4 Boiler failure modes

Failure mode	Mechanism	Appearance	Causes
Overheating	Short-term overheating of tubes leading to final failure by stress rupture.	Longitudinal or 'fish mouth' longitudinal rupture with thin edges.	• Scale or debris blockage in the tube. • Condensate locked in the tube owing to inadequate boil-out.
High-temperature creep	High-temperature failure in the creep temperature range >400 °C.	Blister or larger 'fish-mouth' longitudinal rupture with thick edges. Area around the rupture can have scaly appearance.	• Excessive gas temperatures or over-firing during start-up. • Hot gas flowing through an area of low circulation due to plugging or scaling, etc. • Heat transfer from an adjacent uncooled component (hanger plate, etc.).
Caustic embrittlement	Intergranular attack along grain boundaries, causing cracking.	Common on tubes rolled into vessel shells. General cracking with very little metal loss.	• Excessive gas temperatures for the grade of steel used. • high stresses; • free caustic salts; • a concentrating mechanism.
Caustic gouging	Waterside corrosion that attacks the protective magnetite film on metal surfaces.	Common on the waterside of dirty boiler tubing as either small pinhole leaks or small bulges on thinned tubes, with thin-edged rupture failures.	A combination of: • high heat flux; • caustic contaminants; • inconsistent boiler water chemistry.

Table 10.4 Cont.

Pitting corrosion	Waterside corrosion.	Generally localized, particularly on the inside of tubes. Sharp-edged craters surrounded by red/brown oxide deposits.	Occurs in any areas where oxygen comes into contact with steel. Common causes are: • high O_2 levels in feedwater; • condensate and air remaining in the boiler when shut down.
Stress corrosion cracking	Waterside corrosion most commonly starting on the inside of tubes, resulting in cracks and leakage.	Longitudinal or circumferential thick-edged cracks. The cracks often have a bifurcated (branched) appearance under magnification.	Combination of the presence of: • stress; • cyclic operation; • caustic contaminants.
Corrosion fatigue	Cracking failure of stressed components.	Thick-edged parallell surface cracks with oxide-coating and pits on the fracture surface.	A combination of: • induced stress from a constrained joint or inaccurate assembly; • a corrosive environment.
Hydrogen attack	Waterside corrosion of dirty steam generating tube-banks in high-pressure boilers.	'Window' piece of tube falls out, leaving a thick-edged fracture, without any preliminary wastage or thinning of the tube.	A combination of: • acidic boiler water; • dirty tube surfaces; • high heat flux.
Waterwall corrosion	A general fireside corrosion on the waterwall tubes in a boiler's combustion zone.	General thinning of the tube material accompanied by deep longitudinal cracks and gouges. Hard slag deposits on the fireside surface of the tube.	Too little oxygen in the combustion. Excessive sulphides or chlorides in the fuel.

Exfoliation	'Flaking' of tube and piping surfaces.	Wastage or 'spalling' of the component from its inside surface.	• Constituents of the metal oxide, resulting in differential expansion between the oxide layer and the metal surface. • Quenching of tube internals during transient operations.
Low-temperature corrosion	Common fireside corrosion mechanism on boiler tubes.	Gouging of the external surface and thin edged ductile failure, often in the form of a hole.	Acidic contaminants (such as sulphur or ash) in the furnace gas.

10.6 Failure-related terminology

Acid A compound that yields hydrogen ions (H^+) when dissolved in water.

Alkaline (1) Having properties of an alkali. (2) Having a pH greater than 7.

Alloy steel Steel containing specified quantities of alloying elements added to effect changes in mechanical or physical properties.

Amphoteric Capable of reacting chemically either as an acid or a base. In reference to certain metals, signifies their tendency to corrode at both high and low pH.

Anode In a corrosion cell, the area over which corrosion occurs and metal ions enter solution; oxidation is the principal reaction.

Austenite A face-centred cubic solid solution of carbon or other elements in non-magnetic iron.

Austenitic stainless steel A non-magnetic stainless steel possessing a micro-structure of austenite. In addition to chromium, these steels commonly contain at least 8 per cent nickel.

Blowdown In connection with boilers, the process of discharging a significant portion of the aqueous solution in order to remove accumulated salts, deposits, and other impurities.

Brittle fracture Separation of a solid accompanied by little or no macroscopic plastic deformation.

Cathode In a corrosion cell, the area over which reduction is the principal reaction. It is usually an area that is not attacked.

Caustic cracking A form of stress-corrosion cracking affecting carbon steels and austenitic stainless steels when exposed to concentrated caustic, i.e. high-alkaline solutions.

Caustic embrittlement Term denoting a form of stress corrosion cracking most frequently encountered in carbon steels or iron–chromium–nickel alloys that are exposed to concentrated hydroxide solutions at temperatures of 200–250 °C (400–480 °F).

Cavitation The formation and instantaneous collapse of innumerable tiny voids or cavities within a liquid subjected to rapid and intense pressure changes.

Cavitation damage The degradation of a solid body resulting from its exposure to cavitation. This may include loss of material, surface deformation, or changes in properties or appearance.

Cementite A compound of iron and carbon, known chemically as iron carbide and having the approximate chemical formula Fe_3C.

Cold work Permanent deformation of a metal produced by an external force.

Corrosion fatigue The process in which a metal fractures prematurely under conditions of simultaneous corrosion and repeated cyclic loading. Fracture occurs at lower stress levels or fewer cycles than would be required in the absence of the corrosive environment.

Corrosion product Substance formed as a result of corrosion.

Creep Time-dependent deformation occurring under stress and high temperature.

Creep rupture See **Stress rupture**.

De-alloying (see also **Selective leaching**) The selective corrosion of one or more components of a solid solution alloy. Also called 'parting' or 'selective leaching'.

De-nickelification Corrosion in which nickel is selectively leached from nickel-containing alloys. Most commonly observed in copper–nickel alloys after extended service in fresh water.

De-zincification Corrosion in which zinc is selectively leached from zinc-containing alloys. Most commonly found in copper–zinc alloys containing less than 85 per cent copper after extended service in water containing dissolved oxygen.

Downcomer Boiler tubes in which fluid flow is away from the steam drum.

Ductile fracture Fracture characterized by tearing of metal accompanied by appreciable gross plastic deformation and expenditure of considerable energy.

Ductility The ability of a material to deform plastically without fracturing.

Erosion Destruction of metals or other materials by the abrasive action of moving fluids, usually accelerated by the presence of solid particles or matter in suspension. When corrosion occurs simultaneously, the term 'erosion corrosion' is often used.

Eutectic structure The microstructure resulting from the freezing of liquid metal such that two or more distinct, solid phases are formed.

Exfoliation A type of corrosion that progresses approximately parallel to the outer surface of the metal, causing layers of the metal or its oxide to be elevated by the formation of corrosion products.

Failure A general term used to imply that a part in service (1) has become completely inoperable, (2) is still operable but is incapable of satisfactorily performing its intended function, or (3) has deteriorated seriously, to the point that it has become unreliable or unsafe for continued use.

Fatigue The phenomenon leading to fracture under repeated or fluctuating mechanical stresses having a maximum value less than the tensile strength of the material.

Ferritic stainless steel A magnetic stainless steel possessing a microstructure of alpha ferrite. Its chromium content varies from 11.5 to 27 per cent but it contains no nickel.

Fish-mouth rupture A thin- or thick-lipped burst in a boiler tube that resembles the open mouth of a fish.

Gas porosity Fine holes or pores within a metal that are caused by entrapped gas or by evolution of dissolved gas during solidification.

Grain boundary A narrow zone in a metal corresponding to the transition from one crystallographic orientation to another, thus separating one grain from another.

Graphitic corrosion Corrosion of gray iron in which the iron matrix is selectively leached away, leaving a porous mass of graphite behind. Graphitic corrosion occurs in relatively mild aqueous solutions and on buried pipe fittings.

Graphitization A metallurgical term describing the formation of graphite in iron or steel, usually from decomposition of iron carbide at elevated temperatures.

Heat-affected zone In welding, that portion of the base metal that was not melted during welding, but whose microstructure and mechanical properties were altered by the heat.

Inclusions Particles of foreign material in a metallic matrix. The particles are usually compounds but may be any substance that is foreign to (and essentially insoluble in) the matrix.

Intergranular corrosion Corrosion occurring preferentially at grain boundaries, usually with a slight or negligible attack on the adjacent grains.

Lap A surface imperfection having the appearance of a seam, and caused by hot metal, fins, or sharp corners being folded over and then being rolled or forged into the surface without being welded.

Magnetite A magnetic form of iron oxide, Fe_3O_4. Magnetite is dark gray to black, and forms a protective film on iron surfaces.

Microstructure The structure of a metal as revealed by microscopic examination of the etched surface of a polished specimen.

Overheating Heating of a metal or alloy to such a high temperature that its properties are impaired.

Pitting The formation of small, sharp cavities in a metal surface by corrosion.

Residual stress Stresses that remain within a body as a result of plastic deformation.

Root crack A crack in either a weld or the heat-affected zone at the root of a weld.

Scaling The formation at high temperatures of thick layers of corrosion product on a metal surface.

Scaling temperature A temperature or range of temperatures at which the resistance of a metal to thermal corrosion breaks down.

Selective leaching Corrosion in which one element is preferentially removed from an alloy, leaving a residue (often porous) of the elements that are more resistant to the particular environment.

Spalling The cracking and flaking of particles out of a surface.

Stress corrosion cracking Failure by cracking under combined action of corrosion and stress, either external (applied) stress or internal (residual) stress. Cracking may be either intergranular or transgranular, depending on the metal and the corrosive medium.

Stress raisers Changes in contour or discontinuities in structure that cause local increases in stress.

Stress rupture (creep rupture) A fracture that results from creep.

Tuberculation The formation of localized corrosion products in the form of knob-like mounds called 'tubercles'.

Underbead crack A subsurface crack in the base metal near a weld.

CHAPTER 11

Pressure Equipment: Directives and Legislation

11.1 Introduction: what's this all about?

It is all about compliance. Unlike many types of engineering products, the design, manufacture, and operation of pressure equipment is heavily affected by a mass of legislation and regulations. Whereas in the past these were predominantly national requirements, the current trend to European integration has led to a situation where the objective is to make the requirements pan-European.

11.1.1 The driving forces

There are two main driving forces.
- The objective of the European Union is to eliminate (or at least minimize) barriers to trade between EU member states. This means, essentially, that the requirements governing pressure equipment should be the same in all member states. The mechanism is to implement this by issuing European **Directives**.
- The UK (and all other EU member states) are bound by law to comply with European directives. They do this by reflecting the requirements of the directives in their own national legislation, in the form of **Statutory Instruments**. The requirements of these are, in turn, reflected in the content of various **Regulations**. In the UK the body charged with implementation is the Department of Trade and Industry (DTI) in the form of its Standards and Technical Regulations Directorate (STRD)

11.1.2 The EU 'new approaches'

Pressure equipment legislation has been influenced by two European philosophies designed, nominally, to encourage the free circulation of products within the EU. These are known under their titles of:
- the new approach to product regulation ('the *new approach*')

and
- the global approach to conformity assessment ('the *global approach*')

The idea of 'the new approach' is that products manufactured legally in one EU country should be able to be sold and used freely in the others, without undue restriction. One important factor in this is the idea of technical harmonization of important aspects of the products, known as 'essential requirements'. This means that there is a distinction between the essential requirements of a product and its full technical specification.

The idea of 'the global approach' is that procedures should be in place for reliable conformity assessment of products covered by directives and that this be carried out using the principles of 'confidence through competence and transparency' and putting in place a framework for conformity assessment. Some elements of this are:
- dividing conformity assessment procedures into 'modules' (see later for those relevant to pressure equipment);
- the designation of bodies operating 'conformity assessment procedures' (called 'Notified Bodies');
- the use of the 'CE mark';
- implementing quality assurance and accreditation systems (such as the EN 45000 and EN ISO 9000 series).

Since 1987, various European directives have gradually come into force, based on the philosophy of these two approaches. Several, with direct relevance to pressure equipment, have already been issued and others are in preparation.

11.2 The role of technical standards

Traditionally, most EU countries had (and in most cases still have) their own well-established product standards for all manner of manufactured products, including pressure equipment. Inevitably, these standards differ in their technical and administrative requirements, and often in the fundamental way that *compliance* of products with the standards is assured.

11.2.1 Harmonized standards

Harmonized standards are European standards produced (in consultation with member states) by the European standards organizations CEN/CENELEC. There is a Directive 98/34/EC, which explains the formal status of these harmonized standards. Harmonized standards have to be 'transposed' by each EU country. This means that they must be made available as national standards and that any conflicting standards have to be withdrawn within a given time period.

A key point about harmonized standards is that any product that complies with the standards is automatically assumed to conform to the 'essential requirements' (see later) of the 'new approach' European directive relevant to the particular product. This is known as the 'presumption of conformity'. Once a national standard is transposed from a harmonized standard, then the presumption of conformity is carried with it.

Note that the following terms appear in various Directives, guidance notes, etc. They are all exactly the same thing.
- Essential Safety Requirements (ESRs).
- Essential Requirements.
- Essential Health and Safety Requirements (EHSRs).

Compliance with a harmonized standard is not compulsory; it is voluntary, but compliance with it does infer that a product meets the essential safety requirements (ESRs) of a relevant directive and the product can then carry the 'CE mark'.

11.2.2 National standards

EU countries are at liberty to keep their national product standards if they wish. Products manufactured to these do not, however, carry the 'presumption of conformity' with relevant Directives, hence the onus is on the manufacturer to prove compliance with the ESRs to a Notified Body (on a case-by-case basis). Once compliance has been demonstrated, then the product can carry the CE mark.

11.2.3 The situation for pressure equipment

Pressure equipment is 'a product' so is subject to the type of controls exercised over products as described above. The situation is still developing (and will continue to do so), but Fig. 11.1 broadly summarizes the current situation. Partly for historical reasons, as well as technical ones, vessels that qualify as being 'simple' have their own Simple Pressure Vessels (SPV) Directive, published in 1987. Following the full implementation of The Pressure Equipment Directive in 2002, it is the intention that the SPV Directive will be revised or withdrawn. Note that all the requirements shown in Fig. 11.1 refer to the design and manufacture of pressure equipment – they have no relevance to the inspection of pressure equipment during its working life in the UK [this is addressed by The Pressure System Safety Regulations 2000 (PSSRs), which apply to the UK only].

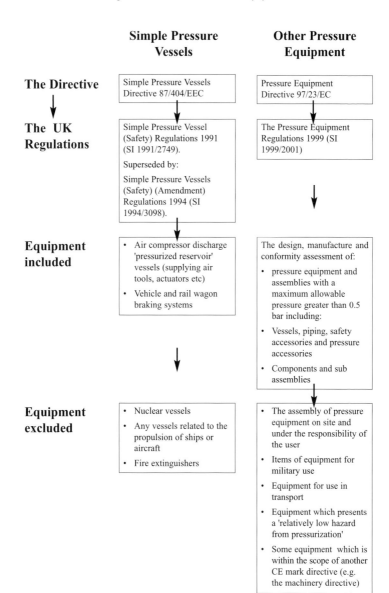

Fig. 11.1 Pressure equipment legislation – the current situation

11.3 Vessel 'statutory' certification

Currently, as the Pressure Equipment Directive (PED) does not have to be complied with until May 2002, the situation in the UK remains in transition. The old (but still existing) system is one of vessel 'statutory certification'. As it will be replaced by the provisions of the directives, the situation is described here in the past tense.

11.3.1 Why was certification needed?

There are four possible reasons why a pressure vessel previously needed to be 'certificated':
- the need for certification was imposed, or inferred, by statutory legislation in the country where the vessel was intended to be installed and used;
- the need for certification was imposed, or inferred, by statutory legislation in the country where the vessel was manufactured;
- the need for certification was imposed, or inferred, by the company that would provide an insurance policy for the vessel itself and second and third party liabilities when it was in use;
- the manufacturer, contractor or end-user chose to obtain certification because they felt that:
  – it helped maintain a good standard of design and workmanship;
  – it provided evidence to help show that legal requirements for 'due-diligence' and 'duty of care' had been met.

Note that three of these reasons were as a result of certification requirement being imposed (or at least inferred) by an external player, the other being a voluntary decision by one or more of the directly involved parties. The main reason for this was that there were some countries in the world where statutory requirements were (and still are) unclear, sometimes contradictory, or non-existent. The more risk-averse vessel manufacturers and contractors often assumed that certification would be necessary, even if evidence of this requirement was difficult to find.

11.3.2 What was certification?

Certification was an attempt to assure the integrity in a way that was generally accepted by external parties. It used accepted national vessel standards or codes as benchmarks of acceptability and good practice. There was little attempt at harmonization of these standards. Certification did address issues of vessel design, manufacture, and testing, but only insofar as these aspects were imposed explicitly by the relevant standard – not more. Certification was evidence, therefore, of code-compliance. Note that compliance with the ASME code was a special case, because of its statutory implications in the USA (see also Chapter 5).

In order to obtain full certification, the organization that issued the certificate had to comply with the activities raised by the relevant code. These differed slightly between codes, but the basic requirements were the need to:
- perform a quite detailed design appraisal;
- ensure the traceability of the materials of construction;
- witness NDT activities and review the results;
- witness the pressure test;
- monitor the manufacturing process;
- issue a certificate (e.g. BS 5500 'form X' or its equivalent).

11.3.3 Who could certificate vessels?

There were two aspects to this: 'independence' and 'competence'.

Independence

The main national pressure vessel codes, including BS 5500 and TRD, required that vessels, if they were to comply fully with the standard, were certificated by an organization which was independent of the manufacturer. This was generally taken to mean that there should be no direct links in the organizations that would cause commercial (or other) pressures to be imposed on the objectivity of the certificating organization's actions.

Competence

Many European countries (this can be where the vessel was to be either manufactured *or* used) had a restricted list of organizations that were deemed competent to perform the certification of vessels. By definition, these organizations had been assessed, usually by governmental authorities, for both independence and competence. ASME vessels were, once again, a special case.

In the UK, the situation was simply that vessel certification had to be performed by a 'competent' person or organization. There was no unilateral detailed guidance as to what qualified a person or organization as being competent. Competent status would normally only be challenged if there was an accident, failure incident, or negligence claim – then the onus would be on the person or organization claiming the status to prove it. There were various registration schemes in existence by which organizations could try to improve their perceived status as a 'competent body' but none of these were imposed directly by any statutory instrument or regulation. The majority of such schemes involved the organization submitting to audits based on BS EN ISO 9001, sometimes with additional requirements added.

This situation will change completely as a result of the implementation of the PED and related mechanisms such as the CE mark.

11.4 The CE mark – what is it?

CE (see Fig. 11.2) probably stands for Communitee Européen, i.e. the French translation of 'European Community'. It could also represent Conformité Européen. Unfortunately, it is far from certain that whoever invented the mark (possibly a bureaucrat in Brussels) had anything particular in mind other than to create a logo which would be universally recognized in the European Union. Given all the national prejudices about language in the different countries of the EU, any original national identity was probably conveniently forgotten by the time it became 'official'. Hence it is best thought of as simply a convenient logo, without any deeper meaning. In its current context, a product can only have the CE mark fixed to it if it complies with all European Directives pertaining to that type of product.

11.5 Simple pressure vessels

Pressure vessels can be divided broadly into 'simple' vessels and those that have more complex features. The general arrangement of a simple vessel is as shown in Chapter 3 (see Fig. 3.11). Note it has no complicated supports or sections and that the ends are dished, not flat. Currently, SPVs have their own European directives, regulations, and harmonized standards.

Fig. 11.2 The 'CE' mark

11.6 The simple pressure vessels directive and regulations

11.6.1 SPVs – summary

Below is a summary of Directive 87/404/EEC, which is the main one relating to simple pressure vessels.

What does the SPV Directive cover?

The Directive applies to series-produced, unfired pressure vessels of welded construction that are intended to contain air or nitrogen at an internal gauge pressure greater than 0.5 bar.

There are also limits to the maximum working pressure and the minimum working temperature, and it has specific requirements covering the geometry of the design and the materials that can be used.

What is the intention of legislation?

To remove technical barriers to trade by harmonizing the laws of Member States covering the design, manufacture, and conformity assessment and to ensure that only safe vessels are placed on the market.

What are typical SPVs?

- Air compressor discharge 'pressurized reservoir' vessels (supplying air tools, actuators, etc.).
- Vehicle and rail wagon braking systems.

What is the history ?

The SPV Directive (87/404/EEC) was adopted on 25 June 1987 and came into force on 1 July 1990. It was implemented in the United Kingdom by the *Simple Pressure Vessel (Safety) Regulations 1991 (SI 1991/2749)*.

The first amending directive (90/488/EEC) was limited to the introduction of a transition period up to 1 July 1992.

The Directive was then amended for a second time by the relevant parts of the CE Marking Directive (93/68/EEC) which were implemented by the *Simple Pressure Vessels (Safety) (Amendment) Regulations 1991 (SI 1994/3098)*.

What are excluded from the SPV Directive?

- Nuclear vessels.
- Any vessels related to the propulsion of ships or aircraft.
- Fire extinguishers.

What are the current issues?

The SPV Directive was the first 'new approach' directive to be adopted, but it has a narrowly defined scope and prescriptive requirements which are more typical of an 'old-style', technical harmonization directive. The narrow scope means that many pressure vessels are outside the scope and considerable EU discussion has been necessary to provide guidance on borderline cases. In addition, the conformity assessment procedures are not based on the Modules Decision 93/465/EEC and do not contain full quality assurance options as routes to compliance.

For these reasons, the future of the SPV Directive will be reviewed when the Pressure Equipment Directive comes fully into force.

11.6.2 Categories of SPVs

The SPV regulations make different provisions for different categories of vessels, depending on their stored energy, expressed in terms of the product of maximum working pressure in bar and its capacity in litres (PS.V).

- Category A consists of vessels whose PS.V is more than 50 bar.litres, and is divided into:
  - Category A.1 consisting of vessels whose PS.V is more than 3000 but not more than 10 000 bar.litres;
  - Category A.2 consisting of vessels whose PS.V is more than 200 but of more than 3000 bar.litres;
  - Category A.3 consisting of vessels whose PS.V is more than 50 but not more than 200 bar.litres.
- Category B consists of vessels whose PS.V is 50 bar.litres, or less.

Safety requirements

The safety requirements for a vessel in **category A** are that:
1. it meets the essential safety requirements;
2. it has safety clearance;
3. the CE marking and the other specified inscriptions have been affixed;
4. in the case of the supply of a vessel or a relevant assembly, it must be accompanied by the manufacturer's instructions;
   In the case of the taking into service of a vessel or a relevant assembly, the manufacturer or importer must ensure that, at the time of the taking into service, the manufacturer's instructions are made available to all those concerned with the vessel's installation and operation.
5. it is in fact safe.

   The safety requirements for a vessel in **category B** are that:
   - it is manufactured in accordance with engineering practice recognized as sound in an EEA state;
   - it bears the specified inscriptions (but not the CE marking);
   - it is in fact safe.

The SPV Essential Safety Requirements (ESRs) are summarized in Section 11.6.2.1. If, however, the vessel has been made in conformity with one of the harmonized standards shown in Table 11.1, then there is an automatic 'presumption of conformity' with the SPV Directive and Corresponding Regulations.

Safety clearance

A vessel in category A has safety clearance once the approved body has issued a certificate of conformity under the EC verification procedure, or an EC certificate of conformity as part of the EC certificate of conformity procedure. The steps necessary to obtain safety clearance are shown in diagrammatic form in Fig. 11.3.

The **first step** is for the manufacturer, or his authorized representative established in the European Economic Area (EEA), to apply for and obtain from an approved body, before series manufacture commences, an EC certificate of adequacy or an EC-type examination certificate (see guidance notes Annex D for details). Where the vessels are to be manufactured so as to conform with a 'relevant national standard', the applicant may choose which certificate to apply for; where that is not the case, the application must be for a type examination certificate.

The **second step** depends on the category of vessel. In the case of vessels in **category A.1**, after commencing series manufacture, the manufacturer, or his authorized representative established in the EEA, must have the vessels checked by an approved body and obtain a certificate of conformity as part of the EC verification procedure (see guidance notes Annex E for details).

However, in the case of vessels in **category A.2 and A.3** there is a choice:
- the manufacturer, or his authorized representative established in the EEA, must comply with the EC verification procedure;

or
- with the EC certificate of conformity procedure (see guidance notes Annex F for details). In this case the manufacturer (not an authorized representative) must, before series manufacture commences, have the design and manufacturing schedule checked by the approved body which issued the certificate of adequacy, or the type-examination certificate. The approved body will issue an EC certificate of conformity as part of the procedure.

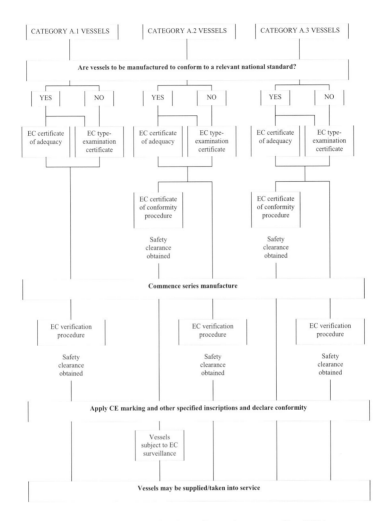

Fig. 11.3 How to obtain safety clearance for SPVs. (Source: Guidance Notes Annex G)

11.6.2.1 SPV essential safety requirements (ESRs)

Part 1 – Materials

1 Materials must be selected according to the intended use of the vessels and in accordance with the following provisions of this Part.

Pressurized components

2 The non-alloy quality steel, non-alloy aluminium, or non-age hardening aluminium alloy used to manufacture the pressurized components must:
   - be capable of being welded;
   - be ductile and tough, so that a rupture at the minimum working temperature does not give rise to either fragmentation or brittle-type fracture;
   - not be adversely affected by ageing.

   For steel vessels, the material must, in addition, meet the requirements set out in paragraph 3 below and, for aluminium or aluminium alloy vessels, those set out in paragraph 4 below. They must be accompanied by an inspection slip.

Steel vessels

3 Non-alloy quality steels must meet the following requirements:
   (a) they must be non-effervescent and be supplied after normalization treatment, or in an equivalent state;
   (b) the content per product of carbon must be less than 0.25 per cent and that of sulphur and phosphorus must each be less than 0.05 per cent;
   (c) they must have the following mechanical properties per product:
   - the maximum tensile strength must be less than 580 N/mm^2;
   - the elongation after rupture must be:
   – if the test piece is taken parallel to the direction of rolling:
      thickness ≥ 3 mm: A ≥ 22%
      thickness < 3 mm: A_{80mm} ≥ 17%
   – if the test piece is taken perpendicular to the direction of rolling:
      thickness ≥ 3 mm: A ≥ 20%
      thickness < 3 mm: A_{80mm} ≥ 15%
   - the average rupture energy for three longitudinal test pieces at the minimum working temperature must not be less than 35 J/cm^2. Not more than one of the three figures may be less than 35 J/cm^2, with a minimum of 25 J/cm^2.

      In the case of steels used to manufacture vessels whose minimum working temperature is lower than -10 °C and whose wall thickness exceeds 5 mm, the average rupture energy must be checked.

Aluminium vessels

4 Non-alloy aluminium must have an aluminium content of at least 99.5 per cent and non-age hardening aluminium alloys must display adequate resistance to intercrystalline corrosion at the maximum working temperature. Moreover, these materials must meet the following requirements:
(a) they must be supplied in an annealed state and
- the maximum tensile strength must be more than 350 N/mm^2;
- the elongation after rupture must be:
  – A ≥ 16% if the test piece is taken parallel to the direction of rolling;
  – A ≥ 14% if the test piece is taken perpendicular to the direction of rolling.

Welding materials

5. The welding materials used to make the welds on or off the vessel must be appropriate to and compatible with the materials to be welded.

Accessories contributing to the strength of the vessel

6. These accessories (bolts, nuts, etc.) must be made either of a material specified in paragraphs 2–4 above or of another kind of steel, aluminium or aluminium alloy which:
- is appropriate to and compatible with the materials used to manufacture the pressurized components;
- at the minimum working temperature has an appropriate elongation after rupture and toughness.

Non-pressurized components

7 All welded non-pressurized components must be of a material which is compatible with that of the parts to which they are welded.

Part 2 – Vessel design

8 The manufacturer must, when designing the vessel, define the use to which it will be put, and select:
- the minimum working temperature;
- the maximum working temperature;
- the maximum working pressure.

However, should a minimum working temperature higher than -10 °C be selected, the properties required of the materials must be satisfied at -10 °C.

The manufacturer must also take account of the following requirements:
- it must be possible to inspect the inside of the vessels;
- it must be possible to drain the vessels;
- the mechanical qualities must be maintained throughout the period of use of the vessel for its intended purpose;
- the vessels must, bearing in mind their envisaged use, be adequately protected against corrosion.

The manufacturer must also be aware of the fact that under the conditions of use envisaged:
- the vessels will not be subjected to stresses likely to impair their safety in use;
- the internal pressure will not permanently exceed the maximum working pressure (however, it may momentarily do so by up to 10 per cent).

Circular and longitudinal seams must be made using full penetration welds or welds of equivalent effectiveness. Dished ends other than hemispherical ones must have a cylindrical edge.

Wall thickness

9   In the case of vessels in category A.2 or A.3, whose maximum working temperature does not exceed 100 °C, the manufacturer must select either the calculation method or the experimental method, as defined below, for determining vessel wall thickness.

In the case of vessels in category A.1, or vessels in Category A.2 or A.3, whose maximum working temperature exceeds 100°C, the calculation method must be used.

However, the actual wall thickness of the cylindrical component and ends must, in any case, not be less than 2 mm in the case of steel vessels, and not less than 3 mm in the case of aluminium or aluminium alloy vessels.

Calculation method

The minimum thickness of the pressurized components must be calculated having regard to the intensity of the stresses and to the following requirements:
- the calculation pressure to be taken into account must not be less than the maximum working pressure;
- the permissible general membrane stress must not exceed 0.6 times the yield strength at the maximum working temperature (RET) or 0.3 times

the tensile strength (R_m), whichever value is the lower. The manufacturer must use the minimum values of RET and R_m guaranteed by the producer of the materials, in order to determine the permissible stress.

However, where the cylindrical component of the vessel has one or more longitudinal welds, made using a non-automatic welding technique, the thickness calculated as above must be multiplied by the coefficient 1.15.

Experimental method

Wall thickness must be determined so as to enable the vessels to resist, at ambient temperature, a pressure equal to at least 5 times the maximum working pressure, with a maximum permanent circumferential deformation factor of 1 per cent.

Part 3 – Manufacturing processes

10 Vessels must be constructed and checked in accordance with the design and manufacturing schedule referred to in Annex D (of the Directive).

Preparation of the component parts

11 The preparation of the component parts (e.g. forming and chamfering) must not give rise to surface defects, cracks, or changes in the mechanical properties of those parts likely to be detrimental to the safety of the vessels.

Welds on pressurized components

12 The characteristics of welds and adjacent zones must be similar to those of the welded materials and must be free of any surface or internal defects detrimental to the safety of the vessels.

Welds must be made by appropriately qualified welders or operators in accordance with approved welding techniques. 'Qualified' means qualified by means of tests carried out by an approved body, and 'approved' means approved by such a body.

The manufacturer must also, during manufacture, ensure consistent weld quality by conducting appropriate tests using adequate procedures. These tests must be the subject of a written report.

Part 4 – Definitions and symbols

Definitions

13 In this Directive:
   (a) *'minimum working temperature'* means the lowest stabilized temperature in the wall of the vessel under normal conditions of use;
   (b) *'inspection slip'* means the document by which the producer of the materials certifies that the materials delivered to the manufacturer meet the requirements set by the manufacturer, and in which the producer sets out the results of the routine inspection tests carried out during the production of those materials (or of materials produced by the same process but not being the materials delivered to the manufacturer) in particular as to their chemical composition and mechanical properties;
   (c) *'maximum working temperature' mean*s the highest stabilized temperature in the wall of the vessel under normal conditions of use:
   (d) *'maximum working pressure'* means the maximum gauge pressure which may be exerted under normal conditions of use;
   (e) *'yield strength at the maximum working temperature'* means:
   - the upper yield point for a material with both a lower and an upper yield point; or
   - the proof stress at 0.2 per cent; or
   - the proof stress at 1.0 per cent in the case of non-alloy aluminium.

Symbols

14 In this Directive:
   (a) 'A' means the percentage elongation after rupture ($L_o = 5.65 \sqrt{S_o}$) where L_o is the gauge length expressed in millimetres and S_o is the cross-sectional area of the test section expressed in square millimetres and 'A_{80mm}' means the percentage elongation after rupture ($L_o = 80$ mm).

11.6.3 SPV harmonized standards

Table 11.1 shows the main harmonized standards for SPVs.

Table 11.1 SPV harmonized standards

Standardization body	Standard reference	Titles *Simple unfired pressure vessels designed to contain air or nitrogen:*	Ratification date
CEN	EN 286-1	Part 1: pressure vessels for general purposes	1998
CEN	EN 286-2	Part 2: pressure vessels for air braking and auxiliary systems for motor vehicles and their trailers	28/09/1992
CEN	EN 286-3	Part 3: steel pressure vessels for air braking equipment and auxiliary pneumatic equipment for railway rolling stock	09/09/1994
CEN	EN 286-4	Part 4: aluminium alloy pressure vessels for air braking equipment and auxiliary pneumatic equipment for railway rolling stock	09/09/1994

11.7 Transportable pressure receptacles: legislation and regulations

Transportable pressure equipment is a generic name given to mainly transportable gas receptacles (TPRs), i.e. gas cylinders, used to contain and transport gases for commercial and domestic purposes. These gases range from nitrogen and oxygen to liquid petroleum gases such as butane and propane.

11.7.1 TPRs legislation

The *Carriage of dangerous goods (classification, packaging and labelling) and use of transportable pressure receptacles regulations* [CDGCPL2] 1996 (S.I.1996/No. 2092) set out requirements for TPRs. CDGCPL2 introduces the ADR and RID European Framework Directives (which are about the carriage of dangerous goods by road or rail) into Great Britain. They contain requirements for:

- design, manufacture, modification, and repair;
- the approval of persons to carry out initial inspection and testing;
- marking and filling.

TPRs (not including EEC-type cylinders, see below) must be manufactured to an HSE approved specification and the manufacture must be verified by a body which is appointed by HSE for this purpose. Tables 11.2 and 11.3 list relevant technical standards.

Table 11.2 Transportable pressure receptable (gas cylinder) technical specifications

Gas cylinder specification	Title, summary, and date	Gas cylinder specification	Title, summary, and date
BS 5045 Part 1	Seamless steel containers: 1982	DOT-4BA (HSE)	Welded steel cylinders made of definitely prescribed steels: 1997
BS 5045 Part 2	Welded steel containers: 1989	HSE-SS-HW3	Hoop wrapped composite seamless steel cylinders: 1999
BS 5045 Part 3	Seamless aluminium containers: 1984	HSE-SS-1000	Welded pressure drums from 900 to 1000 litre capacity manufactured from stainless steel: 1999
LASS 1	Home office lightweight: 1991	DOT-4E (HSE)	Welded aluminium cylinders: 1998
HSE AL HWI	Composite hoop wrapped: 1999	DOT-4L (HSE)	Welded insulated cylinders: 1999
CCL SSGC-1	Chesterfield high strength: 1992	HSE-LHe-TPR	Cryogenic insulated vessels for liquid helium: 1999
HSSS-2	Heiser high strength: 1992	HSE-LL-FW4	Fully wrapped linerless composite cylinder: 1999
HSE-AL-FWI	Fully wrapped composite: 1991	BS EN 1964 Part 1	Seamless steel below R_m 1100 MPa: 1999
DA 2A	Acetylene container (seamless)	BS EN 1975	Seamless aluminium cylinders: 1999

Table 11.2 Cont.

HOAC 1	Acetylene container (welded)	BS EN 1442	Welded steel cylinders for LPG: 1998
MSF-UK	Welded stainless steel gas containers: 1993	BS EN ISO 11120	Gas cylinders – refillable seamless steel tubes of water capacity between 1501 and 3000 litres design, construction and testing: 1999
CPI-3AAX	Large seamless steel containers: 1993	DOT-3A (HSE)	Seamless steel 0.5 – 450 litres, pressure > 10 bar: 2000
BS 5045 Part 5	Welded aluminium alloy containers: 1986	DOT-3B (HSE)	Seamless steel 0.5 –450 litres, pressure 10 – 35 bar, diameter < 127 mm: 2000
CCL-AGC-1	850 N/mm^2 yield strength seamless steel: 1993	DOT-3E (HSE)	Seamless steel > 0.5 litres, pressure > 125 bar, outside diameter <50.8 mm, length < 610 mm: 2000
HSE-AL-FW2	Fully wrapped carbon composite cylinders: 1999	DOT-3BN (HSE)	Seamless nickel 0.5 – 57 litres pressure 10 – 35 bar: 2000
CP1-3T	Large seamless steel (897–1069 N/mm^2) transportable gas containers: 1995	DOT-39 (HSE)	Welded steel non-refillable > 1.4 & < 5 litres, pressure < 155 bar: 2000
HSE-TP-FW3	Fully wrapped carbon/glass fibre composite with thermoplastic liner: 1999	BS EN 1964 Part 3	Seamless stainless steel R_m < 1100 MPa: 2000
HSE-SSGC-1	High-strength seamless steel: 1999	BS EN 1251 Part 2	Vacuum insulated not more than 1000 litres: 2000

Table 11.2 Cont.

DOT-4BW (HSE)	Welded steel cylinders made of definitely prescribed steel with electric-arc welded longitudinal seam: 1995	BS 5045 Part 7	Seamless steel containers (light weight): 2000
FKCO 1120 (HSE)	Welded gas containers from 450 to 1000 litre capacity manufactured from fine grain steel: 1997	BS 5045 Part 8	Seamless aluminium containers (light weight): 2000

Table 11.3 ISO technical standards for gas cylinders

ISO 3807-1: 2000	Cylinders for acetylene; Basic requirements. Part 1: Cylinders without fusible plugs.
ISO 3807-2: 2000	Cylinders for acetylene; Basic requirements. Part 2: Cylinders with fusible plugs.
ISO 4705: 1983	Refillable seamless steel gas cylinders.
ISO 4705: 1983	/Cor 1: 1998.
ISO 4706: 1989	Refillable welded steel gas cylinders.
ISO 5145: 1990	Cylinder valve outlets for gases and gas mixtures; Selection and dimensioning.
ISO 6406: 1992	Periodic inspection and testing of seamless steel gas cylinders.
ISO 7225: 1994	Gas cylinders; Precautionary labels.
ISO/TR 7470: 1988	Valve outlets for gas cylinders; List of provisions which are either standardized or in use.
ISO 7866: 1999	Gas cylinders; Refillable seamless aluminium alloy gas cylinders. Design, construction, and testing.
ISO 9809-1: 1999	Gas cylinders; Refillable seamless steel gas cylinders; Design, construction, and testing. Part 1: Quenched and tempered steel cylinders with tensile strength less than 1100 MPa.
ISO 9809-2: 2000	Gas cylinders; Refillable seamless steel gas cylinders; Design, construction, and testing. Part 2: Quenched and tempered steel cylinders with tensile strength greater than or equal to 1100 MPa.

Table 11.3 Cont.

ISO 10156: 1996	Gases and gas mixtures; Determination of fire potential and oxidizing ability for the selection of cylinder valve outlets.
ISO 10286: 1996	Gas cylinders; Terminology.
ISO 10297: 1999	Gas cylinders; Refillable gas cylinder valves; Specification and type testing.
ISO 10460: 1993	Welded carbon steel gas cylinders; Periodic inspection and testing.
ISO 10461: 1993	Seamless aluminium-alloy gas cylinders; Periodic inspection and testing.
ISO 10462: 1994	Cylinders for dissolved acetylene; Periodic inspection and maintenance.
ISO 10463: 1993	Cylinders for permanent gases; Inspection at time of filling.
ISO 10920: 1997	Gas cylinders; 25E taper thread for connection of valves to gas cylinders; Specification.
ISO 11113: 1995	Cylinders for liquefied gases (excluding acetylene and LPG); Inspection at time of filling.
ISO 11114-1: 1997	Transportable gas cylinders; Compatibility of cylinder and valve materials with gas contents. Part 1: Metallic materials.
ISO 11114-3: 1997	Transportable gas cylinders; Compatibility of cylinder and valve materials with gas contents. Part 3: Autogenous ignition test in oxygen atmosphere.
ISO 11116-1: 1999	Gas cylinders; 17E taper thread for connection of valves to gas cylinders. Part 1: Specifications.
ISO 11116-2: 1999	Gas cylinders – 17E taper thread for connection of valves to gas cylinders – Part 2: Inspection gauges
ISO 11117: 1998	Gas cylinders; Valve protection caps and valve guards for industrial and medical gas cylinders; Design, construction, and tests.
ISO 11118: 1999	Gas cylinders; Non-refillable metallic gas cylinders; Specification and test methods.
ISO 11120: 1999	Gas cylinders; Refillable seamless steel tubes for compressed gas transport, of water capacity between 150 and 3000 litres; Design construction and testing.
ISO 11191: 1997	Gas cylinders; 25E taper thread for connection of valves to gas cylinders; Inspection gauges.
ISO 11372: 1995	Cylinders for dissolved acetylene; Inspection at time of filling.
ISO 11621: 1997	Gas cylinders; Procedures for change of gas service.

Table 11.3 Cont.

ISO 11625: 1998	Gas cylinders; Safe handling.
ISO 11755: 1996	Cylinders in bundles for permanent and liquefiable gases (excluding acetylene); Inspection at time of filling.
ISO 13341: 1997	Transportable gas cylinders; Fitting of valves to gas cylinders.
ISO 13341: 1997	/Cor 1: 1998.
ISO/TR 13763: 1994	Safety and performance criteria for seamless gas cylinders (available in English only).
ISO/TR 13763: 1994	/Cor 1: 1996 (available in English only).
ISO 13770: 1997	Aluminium alloy gas cylinders; Operational requirements for avoidance of neck and shoulder cracks.
ISO/TR 14600: 2000	Gas cylinders; International quality conformance system; Basic rules (available in English only).

The Pressure Vessels (Verification) Regulations 1988 (PVVRs) were intended to implement the European Framework Directive No 76/767/EEC. This set out common provisions for pressure vessels and methods of inspecting them. It also introduced three specific directives that relate to European gas cylinders (i.e. 84/525EEC, 526/EEC, and 527/EEC). The PVVRs set out requirements for the appointment of approved bodies to carry out the inspection of EEC-type cylinders and issue EEC verification certificates. The PVVRs will be superseded by the Pressure Equipment Directive (PED) and Transportable Pressure Equipment Directive (TPED).

Enforcement

In the UK, the Health and Safety Executive (HSE) has general responsibility for the enforcement of the CDGCPL2 and PVVRs. It encourages employers and employees to familiarize themselves with their duties under these regulations if they supply, fill, modify, or repair (re-qualify) TPRs and/or EEC cylinders, or use them as part of work activities. Most workplaces are the responsibility of the HSE Field Operation Directorate, but some fall under the responsibility of the Local Authority, for example offices, shops, and hotels.

The forthcoming TPED

The Transportable Pressure Equipment Directive 99/36/EC (TPED) will be published on 01 July 2001 and will enter into force on 01 July 2003. At this time, it will replace the Simple Pressure Vessels Directive. Below are some key features of the TPED.

- *The approach*

  Unlike the PED, the TPED is not a 'new approach' Directive – it has a similar modular approach (i.e. it has different *construction modules*) but does not specify Essential Safety Requirements (ESRs).

- *It specifies design codes*

  As in other 'old approach' Directives, the TPED specifies design codes (instead of using the ESR approach).

- *Marking*

  Cylinders etc. made to the TPED will be marked with a π (pi) mark, not the 'CE' mark.

- *Scope*

  The TPED contains requirements for in-service inspection as well as new construction (unlike the PED which covers new construction only). It is only relevant to cylinders containing 'class 2' substances (gas), from the nine classes that exist.

- *Enforcement*

  In the UK, the TPED will be enforced by the HSE and DETr (for tankers). A separate system for accreditation of Notified Bodies will be introduced in 2000/20001.

11.8 The pressure equipment directive (PED) 97/23/EC

11.8.1 PED summary

Subject area

The directive covers pressure equipment and assemblies with a maximum allowable pressure PS greater than 0.5 bar. 'Pressure equipment' means vessels, piping, safety accessories, and pressure accessories. 'Assemblies' means several pieces of pressure equipment assembled to form an integrated, functional whole.

Intention of legislation

To remove technical barriers to trade by harmonizing national laws of legislation regarding the design, manufacture, marking, and conformity assessment of pressure equipment.

Coverage

It covers a wide range of equipment such as reaction vessels, pressurized storage containers, heat exchangers, shell and water tube boilers, industrial pipework, safety devices, and pressure accessories. Such equipment is widely used in the chemical, petro-chemical, biochemical, food processing, refrigeration and energy industries, and for power generation.

Implementation

- The Pressure Equipment Regulations 1999 (SI 1999/2001) were laid on 19 July 1999 and came into force on 29 November 1999.
- The Commission's proposal was submitted to the Council of Ministers on 15 July 1993 and was adopted by the European Parliament and the Council on 29 May 1997. It came into force on 29 November 1999, but compliance with its requirements will be optional until 29 May 2002.
- The implementation date for the Directive in all member states is 29 November 1999.
- During a transitional period up to 30 April 2002, member states may permit the placing on the market of any equipment which complies with the legislation in force in that state on 29 May 1997 (the date of adoption of the Directive).

Included

The design, manufacture and conformity assessment of:
- pressure equipment and assemblies with a maximum allowable pressure greater than 0.5 bar including;
- vessels, piping, safety accessories, and pressure accessories;
- components and sub assemblies.

Exclusions

- The assembly of pressure equipment on site and under the responsibility of the user.
- Items of equipment for military use.
- Equipment for use in transport.
- Equipment that presents a 'relatively low hazard from pressurization'.
- Some equipment which is within the scope of another CE mark directive (e.g. the machinery directive).

11.8.2 PED – its purpose

The purpose of the PED is to provide for a legal structure whereby pressure equipment can be manufactured and sold throughout the EU without having to go through a local design approval and inspection regime in every member state. A summary is given in Section 11.8.1. The objective is also to ensure common standards of safety in all pressure equipment sold within the EU, i.e. manufacturers are able to meet the requirements for approval in any member state of the EU, and do not have to repeat the process when selling goods in any other state.

The general idea is that manufacturers will have their equipment approved in their home country. Manufacturers outside of the EU may also have approvals and test work undertaken at their own factory (in many cases this is obligatory) but responsibility for compliance with the requirements of the directive will ultimately rest on the person responsible for selling the product.

Administration

Overall, the requirements of the PED are complex and quite onerous so the principle is that every item of equipment is considered methodically and on an individual basis when trying to decide how best to make it comply with the requirements. The adoption of the new requirements will require changes by everyone involved in the pressure equipment industry. In most cases, however, these changes will be administrative in nature and the actual design of individual pieces of equipment is unlikely to vary very much.

11.8.3 PED – its scope

The PED applies to the design, manufacture, and conformity assessment of pressure equipment and assemblies with a maximum allowable pressure greater than 0.5 bar. Vessels, piping, safety accessories and pressure accessories are all included, see Table 11.4.

Table 11.4 The PED – what does it cover?

Typical equipment covered	Typical equipment excluded
Shell and water tube boilers	The assembly of pressure equipment on site and under the responsibility of the user.
Heat exchangers	Items of equipment for military use.
Plant vessels	Equipment for use in transport.
Pressurized storage containers	Equipment which presents a 'relatively low hazard from pressurization'.
Industrial pipework	Some equipment that is within the scope of another CE mark directive (e.g. the machinery directive).
Gas cylinders	
Certain compressed air equipment	
Safety accessories	
Safety valves,	
Bursting disc safety devices,	
Buckling rods,	
Controlled safety pressure relief systems	
Pressure switches	
Temperature switches	
Fluid level switches	
(Where these are used in safety related applications)	

11.8.4 PED – its structure

The Directive defines a number of classifications for pressure equipment, based on the hazard presented by their application. Hazard is determined on the basis of stored energy (pressure–volume product) and the nature of the contained fluid. Assessment and conformity procedures are different for each category, ranging from self-certification for the lowest (category I) hazard up to full ISO9001 quality management and/or notified body type examination for category IV equipment. Table 11.5 shows the risk categories and their corresponding 'modules'. Figure 11.4 shows the classification and conformity assessment route in graphical form.

Table 11.5 PED risk categories

Risk category	Applicable modules
I	A
II	A1, D1, or E1
III	B1+D, B1+F, B+E, B+C1, H
IV	B+D, B+F, G, or H1

Module	Description (see Sections 11.8.5.1–11.8.5.13 for explanations)
A	Internal production control
A1	Internal production control with monitoring of final assessment
B	EC type examination
B1	EC design examination
C1	Conformity to type
D	Production quality assurance
D1	Production quality assurance
E	Product quality assurance
E1	Product quality assurance
F	Product verification
G	EC unit verification
H	Full quality assurance
H1	Full quality assurance with design examination and special surveillance of final assessment

The assessment procedures are arranged in a modular structure and manufacturers have the choice of which modules to select (within predetermined combinations) in order to best suit their application and manufacturing procedures.

11.8.5 PED – conformity assessment procedures

Sections 11.8.5.1–11.8.5.13 contain a summary of the conformity assessment procedures and responsibilities of the parties contained in Schedule 4 of the Pressure Equipment Regulations (Annex III of the PED). Figure 11.5 reproduces the product classification table and the nine classification charts given in the PED. Refer to the Regulations themselves or *Guidance notes on the UK regulations (URN 99/1147)* for a full statement of the requirements.

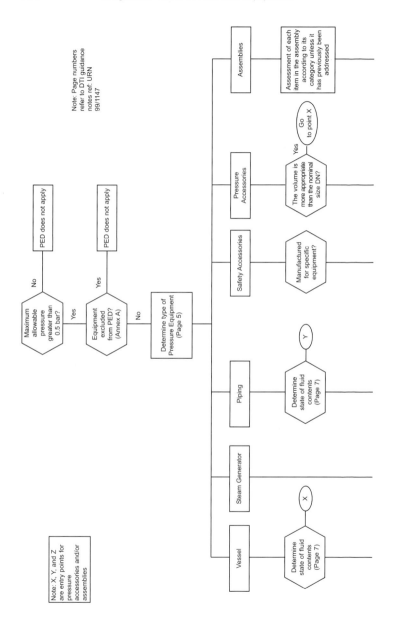

Pressure Equipment: Directives and Legislation

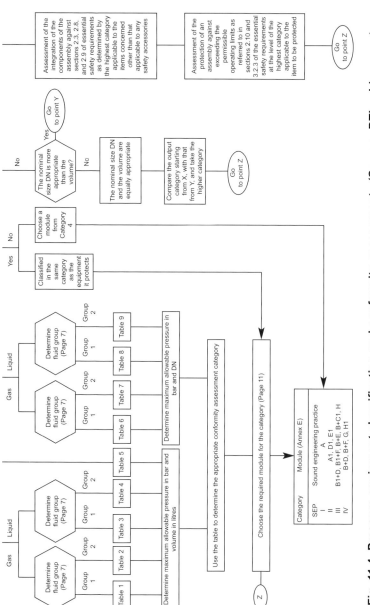

Fig. 11.4 Pressure equipment classification and conformity assessment. (Source: DTI guidance notes on UK Pressure Equipment Regulations: 1999 ref: URN99/1147)

Production classifications

	Vessels				Steam generators	Piping			
State of contents	Gas		Liquid			Gas		Liquid	
Fluid group	1	2	1	2		1	2	1	2
Chart	1	2	3	4	5	6	7	8	9

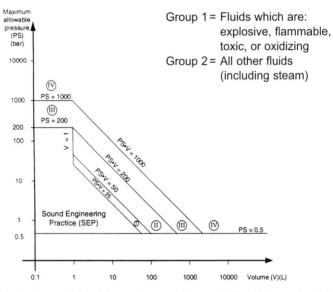

Group 1 = Fluids which are: explosive, flammable, toxic, or oxidizing
Group 2 = All other fluids (including steam)

Exceptionally, vessels intended to contain an unstable gas and falling within categories I or II on the basis of Chart 1 must be classified in category III

Chart 1 Vessels for Group 1 gases

**Fig. 11.5 PED – the nine classification charts.
(Source: DTI guidance notes ref: URN 99/1147)**

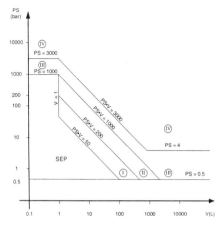

Exceptionally, portable extinguishers and bottles for breathing equipment must be classified at least in category III

Chart 2 Vessels for Group 2 gasses

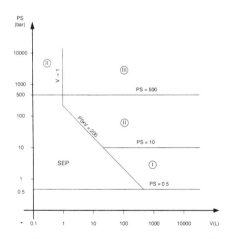

Chart 3 Vessels for Group 1 liquids

Fig. 11.5 PED – the nine classification charts (cont.)

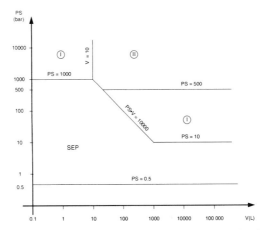

Exceptionally, assemblies intended for generating warm water at temperatures not greater than 110°C which are manually fed with solid fuels and have a product of pressure and volume greater than 50 bar litres, must be subject either to an EC design examination (Module B1) with respect to their conformity with Sections 2.10, 2.11, 3.4, 5(a) and 5(d) of the essential safety requirements, or to full quality assurance (Module H).

Chart 4 Vessels for Group 2 liquids

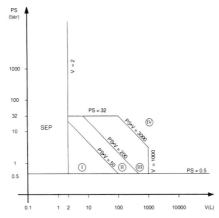

Exceptionally, the design of pressure cookers must be subject to a conformity assessment procedure equivalent to at least one of the category III modules.

Chart 5 Steam generators

Fig. 11.5 PED – the nine classification charts (cont.)

Pressure Equipment: Directives and Legislation 281

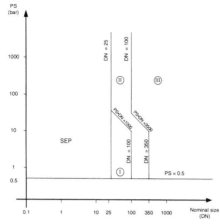

Exceptionally, piping intended for unstable gases and falling within categories I or II must be classified in category III.

Chart 6 Piping for Group 1 gases

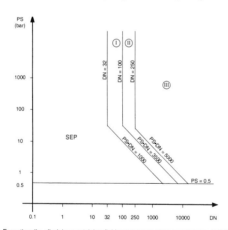

Exceptionally, all piping containing fluids at a temperature greater than 350°C and falling within category II must be classified in category III.

Chart 7 Piping for Group 2 gases

Fig. 11.5 PED – the nine classification charts (cont.)

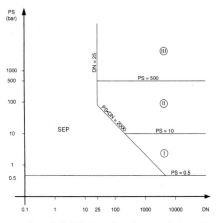

Chart 8 Piping for Group 1 liquids

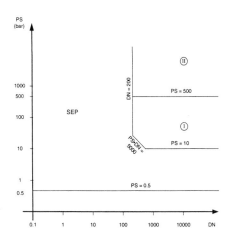

Chart 9 Piping for Group 2 liquids

Fig. 11.5 PED – the nine classification charts (cont.)

11.8.5.1 Module A: Internal production control

Module A describes procedure by which the manufacturer, or his authorized representative established in the Community, ensures and declares that pressure equipment satisfies the requirements of the regulations which apply to it. It does not require the involvement of a notified body.

Manufacturer's duties

- Draw up technical documentation which must enable an assessment of the design, manufacture, and operation of the pressure equipment. It has to contain:
    - a general description of the pressure equipment
    - conceptual design and manufacturing drawings
    - descriptions and explanations necessary for an understanding of the drawings and the operation of the pressure equipment
    - a list of the harmonized standards applied and a description of the solutions adopted to meet the essential safety requirements where harmonized standards have not been applied
    - results design calculations made, examinations carried out etc.
    - test reports
- Ensure manufacturing process complies with technical documentation.
- Affix CE marking.
- Draw up written declaration of conformity.
- Retain declaration of conformity and technical documentation for ten years.

11.8.5.2 Module A1: Internal production control with monitoring of final assessment

In addition to the requirements of Module A:

Manufacturer's duties

- Choose a Notified Body.

Notified Body duties

- Monitor final assessment and monitor by unexpected visits.
- Ensure manufacturer performs final assessment according to Section 3.2 of the essential safety requirements.
- Take samples of pressure equipment at manufacture or storage premises to conduct checks.
- Assess number to sample and whether it is necessary to perform or have performed, all or part of final assessment on the samples.
- Take appropriate action if items do not conform.

Manufacturer's further duties
- Affix Notified Body identification number to each item.

11.8.5.3 *Module B: EC type examination*

Module B describes the part of the procedure where a Notified Body ascertains and attests that a representative example of the production meets the provisions of the regulations which apply to it.

Manufacturer's duties
- Lodge application for EC type examination with a Notified Body which must include:
  - the technical documentation as described under Module A;
  - information on the tests provided for during manufacture;
  - information on the qualifications or approvals of staff carrying out permanent joining and non-destructive tests;
  - information on the operating procedures for permanent joining.
- Provide a representative example of production (a 'type') to the Notified Body.

Notified Body duties
- Examine technical documentation.
- Verify type is manufactured in conformity with the technical documentation.
- Identify parts designed in accordance with harmonized and other relevant standards.
- Assess materials when not in conformity with harmonized standard or European approval and check material certificates.
- Approve procedures for permanent joining or check they have been previously approved.
- Verify that staff are qualified or approved for permanent joining and non-destructive testing.
- Where harmonized standards have been applied, perform appropriate examinations and tests to establish that they have actually been applied.
- Where harmonized standards have not been applied, perform appropriate examinations and tests to establish whether the essential safety requirements have been met.
- If satisfied, issue EC type examination certificate (valid ten years).
- If an EC type examination certificate is refused, provide detailed reasons.
- Additional approval of modifications.
- Retain copies of documentation and EC type examination certificate.

Manufacturer's further duties
- Inform Notified Body of any modifications.
- Retain documentation and copies of EC type examination certificate for ten years.

11.8.5.4 Module B1: EC design examination

Module B1 describes the part of the procedure where a Notified Body ascertains and attests that the design of an item meets the provisions of the regulations which apply to it.

Manufacturer's duties
- Lodge application for EC design examination with Notified Body which must include:
  - technical documentation as described under Module A;
  - supporting evidence for the adequacy of the design solution;
  - information on the qualifications or approvals of staff carrying out permanent joining and non-destructive tests;
  - information on the operating procedures for permanent joining.

Notified Body duties
- Examine technical documentation.
- Identify components that are designed in accordance with harmonized standards and those that are not.
- Assess materials not in conformity with harmonized standards or European approval.
- Approve procedures for permanent joining or check previous approvals.
- Verify staff are qualified or approved for permanent joining and non-destructive testing.
- Where harmonized standards have been applied, perform appropriate examinations to establish that they have been applied properly.
- Where harmonized standards have not been applied, perform appropriate examinations to establish whether the essential safety requirements have been met.
- If satisfied, issue EC design examination certificate.
- If an EC design examination certificate is refused, provide detailed reasons.
- Additional approval of modifications.
- Retain copies of documentation and EC design examination certificate.

Manufacturer's further duties
- Inform Notified Body of modifications to approved design.
- Keep technical documentation and copies of EC design examination certificate for ten years.

11.8.5.5 *Module C1: Monitoring of final assessment*

Module C1 describes procedures where the manufacturer, or authorized representative established in the Community, ensures and declares that the pressure equipment is in conformity with the type as described in the EC type examination certificate (see Module B) and satisfies the requirements of the regulations which apply to it.

Manufacturer's duties

- Ensure that the manufacturing process produces pressure equipment that complies with the type as described in the EC type examination certificate and with the requirements of the regulations.
- Choose a Notified Body.

Notified Body duties

- Monitor final assessment by unexpected visits.
- Ensure manufacturer performs final assessment according to Section 3.2 of the essential safety requirements.
- Take samples of pressure equipment at manufacturing and storage premises to conduct checks.
- Take appropriate action when equipment does not conform.

Manufacturer

- Affix CE marking and Notified Body's identification number.
- Draw up written declaration of conformity.
- Hold copy of declaration of conformity for ten years.

11.8.5.6 *Module D: Quality assurance for production, final inspection, and testing*

Module D describes procedures where the manufacturer ensures and declares that the pressure equipment conforms with the type described in the EC type examination certificate (see Module B) or the EC design examination certificate (see Module B1) and satisfies the requirements of the regulations that apply to it.

Manufacturer's duties

- Operate an approved quality system for production, final inspection, and testing (e.g. ISO 9002) which ensures compliance of the pressure equipment with the type described in the EC type examination certificate or the EC design examination certificate and with the requirements of the regulations that apply to it.

- Lodge application for assessment of quality system with a Notified Body as described under Module D1 below, but with the addition of technical documentation for the approved type and a copy of the EC type examination certificate or EC design examination certificate.
- Undertake to fulfil obligations arising from the quality system.

Notified Body duties

- Assess quality system including an inspection visit to the manufacturer's premises.
- Include in the auditing team at least one member with experience of assessing the pressure equipment technology.
- Presume conformity in respect of the elements of the quality system which implement a relevant harmonized standard (e.g. ISO 9002).
- Notify the manufacturer of assessment decision.
- Carry out surveillance visits to ensure that the manufacturer fulfils the obligations arising from the approved quality system.
- Carry out periodic audits such that a full reassessment is carried out every three years.
- Carry out unexpected visits to verify that the quality system is functioning correctly.
- Assess proposed changes to the quality system.

Manufacturer's further duties

- Affix CE marking and identification number of Notified Body responsible for surveillance.
- Draw up written declaration of conformity.
- Inform the Notified Body of intended adjustments to the quality system.
- Hold documentation for ten years.

11.8.5.7 Module D1: Quality assurance for production, final inspection, and testing

Module D1 describes procedures where the manufacturer ensures and declares that the items of pressure equipment satisfy the requirements of the regulations that apply to them.

Manufacturer's duties

- Draw up technical documentation covering design, manufacture, and operation as described under Module A.
- Operate an approved quality system for production, final inspection, and testing (e.g. ISO 9002) which must ensure compliance of the pressure equipment with the requirements of the regulations that apply to it.
- Lodge application for assessment of quality system with a Notified Body

which includes:
- relevant information on the pressure equipment concerned;
- documentation on the quality system including a description of quality objectives and organizational structure;
- manufacturing, quality control, and quality assurance techniques to be used;
- examinations and tests to be carried out;
- quality records, such as inspection reports and test data, calibration data, reports concerning the qualifications or approvals of the personnel concerned, particularly with permanent joining;
- means of monitoring quality and the quality system.
- Undertake to fulfil obligations arising from the quality system.

Notified Body duties
- Assess quality system including an inspection visit to the manufacturer's premises.
- Include in the auditing team at least one member with experience of assessing the pressure equipment technology.
- Presume conformity in respect of the elements of the quality system which implement a relevant harmonized standard (e.g. ISO 9002).
- Notify the manufacturer of the assessment decision.
- Carry out surveillance visits to ensure that the manufacturer fulfils the obligations arising from the approved quality system.
- Carry out periodic audits such that a full reassessment is carried out every three years.
- Carry out unexpected visits to verify that the quality system is functioning correctly.
- Assess proposed changes to the quality system.

Manufacturer's further duties
- Affix CE marking and identification number of Notified Body responsible for surveillance.
- Draw up written declaration of conformity.
- Inform the Notified Body of intended adjustments to the quality system.
- Hold documentation for ten years.

NB: For category III and IV vessels for Group 1 gases, Group 2 gases and Group 1 liquids and steam generators, the Notified Body, when performing unexpected visits, must take a sample of equipment and perform or have performed, the proof test referred to in Section 3.2.2 of the essential safety requirements.

11.8.5.8 Module E: Quality assurance for final inspection and testing

Module E describes the procedures where the manufacturer ensures and declares the equipment is in conformity with the type described in the EC type examination certificate (see Module B) and satisfies the requirements of the regulations that apply to it.

Manufacturer's duties

- Operate an approved quality system for production, final inspection, and testing (e.g. ISO 9003) under which each item of pressure equipment must be examined and appropriate tests carried out to ensure its conformity with the requirements of the regulations that apply to it.
- Lodge application for assessment of quality system with a Notified Body as described under Module E1 below but with the addition of technical documentation for the approved type and a copy of the EC type examination certificate or EC design examination certificate.
- Undertake to fulfil obligations arising from the quality system.
- Examine and test equipment as set out in relevant harmonized standard, or equivalent and, particularly, carry out final assessment as referred to in Section 3.2 of the essential safety requirements.

Notified Body duties

- Assess quality system including an inspection visit to the manufacturer's premises.
- Include in the auditing team at least one member with experience of assessing the pressure equipment technology.
- Presume conformity in respect of the elements of the quality system which implement a relevant harmonized standard (e.g. ISO 9003).
- Notify the manufacturer of the assessment decision.
- Carry out surveillance visits to ensure that the manufacturer fulfils the obligations arising from the approved quality system.
- Carry out periodic audits such that a full reassessment is carried out every three years.
- Carry out unexpected visits to verify that the quality system is functioning correctly.
- Assess proposed changes to the quality system.

Manufacturer's further duties

- Affix CE marking and identification number of Notified Body responsible for surveillance.
- Draw up written declaration of conformity.
- Inform the Notified Body of intended adjustments to the quality system.
- Hold documentation for ten years.

11.8.5.9 *Module E1: Quality assurance for final inspection and testing*

Module E1 describes the procedure where the manufacturer ensures and declares that the equipment satisfies the requirements of the regulations that apply to it.

Manufacturer's duties

- Draw up technical documentation covering design, manufacture, and operation as described under Module A.
- Operate an approved quality system for production, final inspection, and testing (e.g. ISO 9003) under which each item of pressure equipment must be examined and appropriate tests carried out to ensure its conformity with the requirements of the regulations that apply to it.
- Lodge application for assessment of quality system with a Notified Body which includes:
  - relevant information on the pressure equipment concerned
  - documentation on the quality system including a description of quality objectives and organizational structure;
  - procedures for the permanent joining of parts;
  - examinations and tests to be carried out after manufacture;
  - quality records, such as inspection reports and test data, calibration data, reports concerning the qualifications or approvals of the personnel concerned, particularly with permanent joining;
  - means of monitoring quality and the quality system.
- Undertake to discharge obligations arising from the quality system.

Notified Body duties

- Assess quality system including an inspection visit to the manufacturer's premises.
- Include in the auditing team at least one member with experience of assessing the pressure equipment technology.
- Presume conformity in respect of the elements of the quality system which implement a relevant harmonized standard (e.g. ISO 9002).
- Notify the manufacturer of the assessment decision.
- Carry out surveillance visits to ensure that the manufacturer fulfils the obligations arising from the approved quality system.
- Carry out periodic audits such that a full reassessment is carried out every three years.
- Carry out unexpected visits to verify that the quality system is functioning correctly.
- Assess proposed changes to the quality system.

Manufacturer's further duties
- Affix CE marking and identification number of Notified Body responsible for surveillance.
- Draw up written declaration of conformity.
- Inform the Notified Body of intended adjustments to the quality system.
- Hold documentation for ten years.

11.8.5.10 Module F: Product verification

Module F describes the procedure where the manufacturer, or his authorized representative established in the Community, ensures and declares the pressure equipment is in conformity with the type as described in the EC type examination certificate or the EC design examination certificate and satisfies the requirements of the regulations that apply to it.

Manufacturer's duties
- Ensure that the manufacturing process produces pressure equipment that is in conformity with the type described in the EC type examination certificate (see Module B) or EC design examination certificate (see Module B1) and the requirements of the regulations which apply to it.
- Choose a Notified Body.

Notified Body duties
- Examine and test each item of pressure equipment, as set out in the relevant harmonized standard or equivalent, to verify that every item conforms to the type and the requirements of the regulations.
- Verify qualifications of personnel responsible for permanent joining of parts and non-destructive examination.
- Verify material manufacturers' certificates.
- Carry out, or have carried out, final inspection and proof test as referred to in Section 3.2 of the essential safety requirements.
- Examine safety devices, if applicable.
- Draw up written certificate of conformity relating to the tests.
- Affix, or have affixed, its identification number to each item.

Manufacturer's further duties
- Affix the CE marking.
- Draw up written declaration of conformity.
- Ensure that the certificates of conformity issued by the Notified Body are available on request.
- Keep copy of the declaration of conformity for ten years.

11.8.5.11 Module G: Unit verification

Module G describes the procedure where the manufacturer ensures and declares the pressure equipment that has been issued with a certificate of conformity for tests carried out satisfies the requirements of the regulations which apply to it.

Manufacturer's duties

- lodge application for unit verification with a Notified Body which must include:
  - the technical documentation, as described under Module A;
  - information relating to the approval of the manufacturing and test procedures;
  - information on the qualifications or approvals of staff carrying out permanent joining and non-destructive tests;
  - information on the operating procedures for permanent joining.

Notified Body duties

- Examine design and construction of each item.
- Examine technical documentation with respect to design and manufacturing procedures.
- Assess materials used where they do not conform to a relevant harmonized standard or European materials approval.
- Check certificates from material manufacturers.
- Approve procedures for permanent joining of parts to check previous approval.
- Verify personnel responsible for permanent joining of parts and non-destructive testing.
- Perform, during manufacture, appropriate tests set out in relevant harmonized standards or equivalent to ensure conformity with regulations.
- Carry out final inspection and perform, or have performed, proof test.
- Examine the safety devices, if applicable.
- Affix identification number or have it affixed to the pressure equipment.
- Draw up certificate of conformity for the tests carried out.

Manufacturer's further duties

- Affix CE marking.
- Draw up written declaration of conformity.
- Ensure that the certificates of conformity and declaration of conformity are available on request.

11.8.5.12 Module H: Full quality assurance

Module H describes the procedure where the manufacturer ensures and declares that the pressure equipment satisfies the requirements of the regulations that apply to it.

Manufacturer's duties

- Implement an approved quality system for design, manufacture, and testing (e.g. ISO 9001) which must ensure compliance of the pressure equipment with the requirements of the regulations that apply to it.
- Lodge application for assessment of quality system with a Notified Body which includes:
  - relevant information on the pressure equipment concerned;
  - documentation on the quality system including a description of:
    - quality objectives and organizational structure;
    - technical design specifications, including standards, that will be applied;
    - design control and verification techniques, processes and systematic measures, particularly with regard to materials;
    - manufacturing quality control and quality assurance techniques to be used, particularly the procedures for permanent joining;
    - examinations and tests to be carried out;
    - quality records, such as inspection reports and test data, calibration data, reports concerning the qualifications or approvals of the personnel concerned, particularly with permanent joining;
    - means of monitoring quality and the quality system.
- Undertake to fulfil obligations arising from the quality system.

Notified Body duties

- Assess quality system, including an inspection visit to the manufacturer's premises.
- Include in the auditing team at least one member with experience of assessing the pressure equipment technology.
- Presume conformity in respect of the elements of the quality system which implement a relevant harmonized standard (e.g. ISO 9001).
- Notify the manufacturer of the assessment decision.
- Carry out surveillance visits to ensure that the manufacturer fulfils the obligations arising from the approved quality system.
- Carry out periodic audits such that a full reassessment is carried out every three years.
- Carry out unexpected visits to verify that the quality system is functioning correctly.
- Assess proposed changes to the quality system.

Manufacturer's further duties
- Affix CE marking and identification number of Notified Body responsible for surveillance.
- Draw up written declaration of conformity.
- Inform the Notified Body of intended adjustments to the quality system.
- Hold documentation for ten years.

11.8.5.13 Module H1: Full quality assurance with Design examination and monitoring of the final assessment

In addition to the requirements of Module H

Manufacturer's duties
- Lodge application for examination of the design with a Notified Body. The application must enable the design, manufacture, and operation of the pressure equipment to be understood and enable conformity with the relevant requirements for the regulations to be assessed. It must include:
  – technical design specifications, including standards;
  – necessary supporting evidence for their adequacy, in particular where harmonized standards have not been applied in full;
  – inform the Notified Body of all modifications to the approved design

Notified Body duties
- Examine the application and, if satisfied, issue an EC design examination certificate.
- Carry out increased surveillance of the final assessment in the form of unexpected visits which must include examinations on the pressure equipment.
- Assess any modifications to the approved design and give additional approval.

11.8.6 Essential safety requirements (ESRs)

The PED defines certain requirements about the design and performance of pressure equipment in a series of Essential Safety Requirements (ESRs), contained in Annex 1 of the Directive itself. These are compulsory. Two important clauses from the PED are:

Clause 3: *The manufacturer is under an obligation to analyse the hazards in order to identify those which apply to his equipment on account of pressure; he must then design and construct it taking account of his analysis.*

Clause 4 : *The essential requirements are to be interpreted and applied in such a way as to take account of the state of the art and current practice at the time of design and manufacture as well as of technical and economic considerations which are consistent with a high degree of health and safety protection.*

Section 11.8.6.1 summarizes the situation. Table 4.5 (Chapter 4) shows how PD 5500 (formerly BS 5500) claims compliance with the ESRs.

11.8.6.1 PED essential safety requirements – checklist

This section gives the Essential Safety Requirements (ESRs) required for compliance with the PED. The ESR numbering is as the PED document. The ESRs are introduced by the following four 'comments':

- The obligations arising from the essential requirements listed in this Annex for pressure equipment also apply to assemblies where the corresponding hazard exists.
- The essential requirements laid down in the Directive are compulsory. The obligations laid down in these essential requirements apply only if the corresponding hazard exists for the pressure equipment in question when it is used under conditions that are reasonably foreseeable by the manufacturer.
- The manufacturer is under an obligation to analyse the hazards in order to identify those which apply to his equipment on account of pressure; he must then design and construct it, taking account of his analysis.
- The essential requirements are to be interpreted and applied in such a way as to take account of the state of the art and current practice at the time of design and manufacture as well as of technical and economic considerations which are consistent with a high degree of health and safety protection.

ESRs checklist

1 General

   1.1 Pressure equipment must be designed, manufactured, and checked. If applicable it must be equipped and installed in such a way as to ensure its safety when put into service in accordance with the manufacturer's instructions, or in reasonably foreseeable conditions.

   1.2 In choosing the most appropriate solutions, the manufacturer must apply the principles set out below in the following order:
   – eliminate or reduce hazards as far as is reasonably practicable;

- apply appropriate protection measures against hazards that cannot be eliminated;
- where appropriate, inform users of residual hazards and indicate whether it is necessary to take appropriate special measures to reduce the risks at the time of installation and/or use.

1.3 Where the potential for misuse is known or can be clearly foreseen, the pressure equipment must be designed to prevent danger from such misuse. If that is not possible, adequate warning must be given that the pressure equipment must not be used in that way.

2 Design

2.1 General

The pressure equipment must be properly designed, taking all relevant factors into account in order to ensure that the equipment will be safe throughout its intended life.

The design must incorporate appropriate safety coefficients using comprehensive methods which are known to incorporate adequate safety margins against all relevant failure modes in a consistent manner.

2.2 Design for adequate strength

2.2.1 The pressure equipment must be designed for loadings appropriate to its intended use and other reasonably foreseeable operating conditions. In particular, the following factors must be taken into account:
- internal/external pressure;
- ambient and operational temperatures;
- static pressure and mass of contents in operating and test conditions;
- traffic, wind, earthquake loading;
- reaction forces and moments which result from the supports, attachments, piping etc;
- corrosion and erosion, fatigue, etc;
- decomposition of unstable fluids.

Various loadings which can occur at the same time must be considered, taking into account the probability of their simultaneous occurrence.

2.2.2 Design for adequate strength must be based on:
- as a general rule, a calculation method, as described in 2.2.3, and supplement if necessary by an experimental design method as described in 2.2.4; or
- an experimental design method without calculation, as described in 2.2.4 when the product of the maximum allowable pressure PS and the volume V is less than 6000 bar-L or the product PS-DN less than 3000 bar.

2.2.3 Calculation method

(a) Pressure containment and other loading aspects

The allowable stresses for pressure equipment must be limited having regard to reasonably foreseeable failure modes under operating conditions. To this end, safety factors must be applied to eliminate fully any uncertainty arising out of manufacture, actual operational conditions, stresses, calculation models, and the properties and behaviour of the material.

These calculation methods must provide sufficient safety margins consistent, where applicable, with the requirements of Section 7.

The requirements set out above may be met by applying one of the following methods, as appropriate, if necessary as a supplement to or in combination with another method:
- design by formula;
- design by analysis;
- design by fracture mechanics.

(b) Resistance

Appropriate design calculations must be used to establish the resistance of the pressure equipment concerned.

In particular:
- the calculation pressures must not be less than the maximum allowable pressures and take into account static head and dynamic fluid pressures and the decomposition of unstable fluids. Where a vessel is separated into individual pressure-containing chambers, the partition wall must be designed on the basis of the highest possible chamber pressure relative to the lowest pressure possible in the adjoining chamber;

- the calculation temperatures must allow for the appropriate safety margins;
- the design must take appropriate account of all possible combinations of temperature and pressure that might arise under reasonably foreseeable operating conditions for the equipment;
- the maximum stresses and peak stress concentrations must be kept within safe limits;
- the calculation for pressure containment must utilize the values appropriate to the properties of the material, based on documented data, having regard to the provisions set out in Section 4 together with appropriate safety factors. Material characteristics to be considered, where applicable, to include:
    - yield strength, 0.2% or 1.0% proof strength as appropriate at calculation temperature;
    - tensile strength;
    - time-dependent strength, i.e. creep strength;
    - fatigue data;
    - Young's modulus (modulus of elasticity);
    - appropriate amount of plastic strain;
    - impact strength;
    - fracture toughness.
- appropriate joint factors must be applied to the material properties depending, for example on the type of non-destructive testing, the material joined, and the operating conditions envisaged;
- the design must take appropriate account of all reasonably foreseeable degradation mechanisms (e.g. corrosion, creep, fatigue) commensurate with the intended use of the equipment. Attention must be drawn, in the instructions referred to in Section 3.4, to particular features of the design which are relevant to the life of the equipment, for example:
    - for creep: design hours of operation at specified temperatures;
    - for fatigue: design number of cycles at specified stress levels;
    - for corrosion: design corrosion allowance.

(c) Stability aspects

Where the calculated thickness does not allow for adequate structural stability, the necessary measures must be taken to remedy the situation taking into account the risks from transport and handling.

2.2.4 Experimental design method

The design of the equipment may be validated, in all or in part, by an appropriate test programme carried out on a sample representative of the equipment or the category of equipment.

The test programme must be clearly defined prior to testing and accepted by the notified body responsible for the design conformity assessment module, where it exists.

The programme must define test conditions and criteria for acceptance or refusal. The actual values of the essential dimensions and characteristics of the materials which constitute the equipment tested shall be measured before the test.

Where appropriate, during tests, it must be possible to observe the critical zones of the pressure equipment with adequate instrumentation capable of registering strains and stresses with sufficient precision.

The test programme must include:

(a) A pressure strength test, the purpose of which is to check that, at a pressure with a defined safety margin in relation to the maximum allowable pressure, the equipment does not exhibit significant leaks or deformation exceeding a determined threshold.

The test pressure must be determined on the basis of the differences between the values of the geometrical and material characteristics, measured under test conditions and the values used for design purposes; it must take into account the differences between the test and design temperatures.

(b) Where the risk of creep or fatigue exists, appropriate tests determined on the basis of the service conditions laid down for the equipment, for instance hold time at specified temperatures, number of cycles at specified stress levels, etc;

(c) Where necessary, additional tests concerning other factors referred to in 2.2.1 such as corrosion, external damage, etc.

2.3 *Provisions to ensure safe handling and operation*

The method of operation specified for pressure equipment must be such as to preclude any reasonably foreseeable risk in operation of the equipment. Particular attention must be paid, where appropriate, to:
– closures and openings;
– dangerous discharge of pressure relief blow-offs;
– devices to prevent physical access whilst pressure or a vacuum exists;
– surface temperature taking into consideration the intended use;
– decomposition of unstable fluids.

In particular, pressure equipment fitted with an access door must be equipped with an automatic or manual device enabling the user easily to ascertain that the opening will not present any hazard.

Furthermore, where the opening can be operated quickly, the pressure equipment must be fitted with a device to prevent it being opened whenever the pressure or temperature of the fluid presents a hazard.

2.4 *Means of examination*

(a) Pressure equipment must be designed and constructed so that all necessary examinations to ensure safety can be carried out.

(b) Means of determining the internal conditions of the equipment must be available, where it is necessary to ensure the continued safety of the equipment, such as access openings allowing physical access to the inside of the pressure equipment so that appropriate examinations can be carried out safely and ergonomically.

(c) Other means of ensuring the safe condition of the pressure equipment may be applied:
– where it is too small for physical internal access; or
– where opening the pressure equipment would adversely affect the inside; or
– where the substance contained has been shown not to be harmful to the material from which the pressure equipment is made and no other internal degradation mechanisms are reasonably foreseeable.

Pressure Equipment: Directives and Legislation 301

2.5 Means of draining and venting

Adequate means must be provided for the draining and venting of pressure equipment where necessary:
- to avoid harmful effects such as water hammer, vacuum collapse, corrosion, and uncontrolled chemical reactions. All stages of operation and testing, particularly pressure testing, must be considered;
- to permit cleaning, inspection and maintenance in a safe manner.

2.6 Corrosion or other chemical attack

Where necessary, adequate allowance or protection against corrosion or other chemical attack must be provided, taking due account of the intended and reasonably foreseeable use.

2.7 Wear

Where severe conditions of erosion or abrasion may arise, adequate measures must be taken to:
- minimize that effect by appropriate design, e.g. additional material thickness, or by the use of liners or cladding materials;
- permit replacement of parts which are most affected;
- draw attention, in the instructions referred to in 3.4, to measures necessary for continued safe use.

2.8 Assemblies

Assemblies must be designed so that:
- the components to be assembled together are suitable and reliable for their duty;
- all the components are properly integrated and assembled in an appropriate manner.

2.9 Provisions for filling and discharge

Where appropriate, the pressure equipment must be so designed and provided with accessories, or provision made for their fitting, as to ensure safe filling and discharge in particular with respect to hazards such as:

(a) on filling
- overfilling or over pressurization having regard in particular to the filling ratio and to vapour pressure at the reference temperature,
- instability of the pressure equipment;

(b) on discharge: the uncontrolled release of the pressurized fluid;

(c) on filling or discharge: unsafe connection and disconnection.

2.10 Protection against exceeding the allowable limits of pressure equipment

Where, under reasonably foreseeable conditions, the allowable limits could be exceeded, the pressure equipment must be fitted with, or provision made for the fitting of, suitable protective devices, unless the equipment is intended to be protected by other protective devices within an assembly.

The suitable device or combination of such devices must be determined on the basis of the particular characteristics of the equipment of assembly.

Suitable protective devices and combinations thereof comprise:
(a) safety accessories as defined in Article 1, Section 2.1.3;
(b) where appropriate, adequate monitoring devices such as indicators and/or alarms which enable adequate action to be taken either automatically or manually to keep the pressure equipment within the allowable limits.

2.11 Safety accessories

2.11.1 Safety accessories must:
- be so designed and constructed as to be reliable and suitable for their intended duty and take into account the maintenance and testing requirements of the devices, where applicable;
- be independent of other functions, unless their safety function cannot be affected by such other functions;
- comply with appropriate design principles in order to obtain suitable and reliable protection. These principles include, in particular, fail-safe modes, redundancy, diversity, and self-diagnosis.

2.11.2 Pressure limiting devices

These devices must be so designed that the pressure will not permanently exceed the maximum allowable pressure PS; however a short duration pressure surge in keeping with the specifications laid down in 7.3 is allowable, where appropriate.

2.11.3 Temperature-monitoring devices

These devices must have an adequate response time on safety grounds, consistent with the measurement function.

2.12 External fire

Where necessary, pressure equipment must be so designed and, where appropriate, fitted with suitable accessories, or provision made for their fitting, to meet damage-limitation requirements in the event of external fire, having particular regard to its intended use.

3 Manufacturing

3.1 Manufacturing procedures

The manufacturer must ensure the competent execution of the provisions set out at the design stage by applying the appropriate techniques and relevant procedures, especially with a view to the aspects set out below:

3.1.1 Preparation of the component parts

Preparation of the component parts (e.g. forming and chamfering) must not give rise to defects or cracks or changes in the mechanical characteristics likely to be detrimental to the safety of the pressure equipment.

3.1.2 Permanent joining
- Permanent joints and adjacent zones must be free of any surface or internal defects detrimental to the safety of the equipment.
- The properties of permanent joints must meet the minimum properties specified for the materials to be joined unless other relevant property values are specifically taken into account in the design calculations.
- For pressure equipment, permanent joining of components, which contribute to the pressure resistance of equipment and components which are directly attached to them, must be carried out by suitably qualified personnel according to suitable operating procedures.
- For pressure equipment in categories II, III and IV, operating procedures and personnel must be approved by a competent third party which, at the manufacturer's direction, may be:
  - a notified body,
  - a third-party organization recognized by a Member State as provided for in Article 13.

To carry out these approvals the third party must perform examinations and tests as set out in the appropriate harmonized standards or equivalent examinations and tests or must have them performed.

3.1.3 Non-destructive tests

For pressure equipment, non-destructive tests of permanent joints must be carried out by suitable qualified personnel. For pressure equipment in categories III and IV, the personnel must be approved by a third-party organization recognized by a Member State pursuant to Article 13.

3.1.4 Heat treatment

Where there is a risk that the manufacturing process will change the material properties to an extent which would impair the safety of the pressure equipment, suitable heat treatment must be applied at the appropriate stage of manufacture.

3.1.5 Traceability

Suitable procedures must be established and maintained for identifying the material making up the components of the equipment which contribute to pressure resistance by suitable means from receipt, through production, up to the final test of the manufactured pressure equipment.

3.2 Final assessment

Pressure equipment must be subjected to final assessment as described below:

3.2.1 Final inspection

Pressure equipment must undergo a final inspection to assess visually and by examination of the accompanying documents compliance with the requirements of the Directive. Tests carried out during manufacture may be taken into account. As far as is necessary on safety grounds, the final inspection must be carried out internally and externally on every part of the equipment, where appropriate in the course of manufacture (e.g. where examination during the final inspection is no longer possible).

3.2.2 Proof test

Final assessment of pressure equipment must include a test for the pressure containment aspect, which will normally take the form of a hydrostatic pressure test at a pressure at least equal, where appropriate, to the value laid down in 7.4.

For category 1 series-produced pressure equipment, this test may be performed on a statistical basis.

Where hydrostatic pressure test is harmful or impractical, other tests of a recognized value may be carried out. For tests other than hydrostatic pressure test, additional measures, such as non-destructive tests or other methods of equivalent validity, must be applied before those tests are carried out.

3.2.3 Inspection of safety devices

For assemblies, the final assessment must also include a check of the safety devices intended to check full compliance with the requirements referred to in 2.10.

3.3 Marking and labelling

In addition to the CE marking referred to in Article 15, the following information must be provided:

(a) for all pressure equipment:
- the name and address or other means of identification of the manufacturer and, where appropriate, of his authorized representative established within the Community;
- the year of manufacture;
- identification of the pressure equipment according to its nature, such as type, series or batch identification and serial number;
- essential maximum/minimum allowable limits.

(b) depending on the type of pressure equipment, further information necessary for safe installation, operation or use and, where applicable, maintenance and periodic inspection such as:
- the volume V of the pressure equipment in L;
- the nominal size for piping DN;
- the test pressure PT applied in bar and date;
- safety device set pressure in bar;
- output of the pressure equipment in kW;
- supply voltage in V (volts);
- intended use;
- filling ratio kg/L;
- maximum filling mass in kg;
- tare mass in kg;
- the product group.

(c) where necessary, warnings fixed to the pressure equipment drawing attention to misuse which experience has shown might occur.

The CE marking and the required information must be given on the pressure equipment or on a dataplate firmly attached to it, with the following exceptions:
- where applicable, appropriate documentation may be used to avoid repetitive marking of individual parts such as piping components, intended for use with the same assembly. This applies to CE marking and other marking and labelling referred to in this Annex;
- where pressure equipment is too small, e.g. accessories, the information referred to in (b) may be given on a label attached to that pressure equipment;
- labelling or other adequate means may be used for the mass to be filled and the warnings referred to in (c), provided it remains legible for the appropriate period of time.

3.4 Operating instructions

(a) When pressure equipment is placed on the market, it must be accompanied, as far as relevant, with instructions for the user, containing all the necessary safety information relating to:
- mounting including assembling of different pieces of pressure equipment;
- putting into service;
- use;
- maintenance including checks by user.

(b) Instructions must cover information affixed to the pressure equipment in accordance with 3.3, with the exception of serial identification, and must be accompanied, where appropriate, by the technical documents, drawings and diagrams necessary for a full understanding of these instructions;

(c) If appropriate, these instructions must also refer to hazards arising from misuse in accordance with 1.3 and particular features of the design in accordance with 2.2.3.

4 Materials

Materials used for the manufacture of pressure equipment must be suitable for such application during the scheduled lifetime unless replacement is foreseen.

Welding consumables and other joining materials need fulfil only the relevant requirements of 4.1, 4.2 (a) and the first paragraph of 4.3, in an appropriate way, both individually and in a joined structure.

4.1 Materials for pressurized parts must:
  (a) have appropriate properties for all operating conditions which are reasonably foreseeable and for all test conditions, and in particular they should be sufficiently ductile and tough. Where appropriate, the characteristics of the materials must comply with the requirements of 7.5. Moreover, due care should be exercised in particular in selecting materials in order to prevent brittle-type fracture where necessary; where for specific reasons brittle material has to be used, appropriate measures must be taken;
  (b) be sufficiently chemically resistant to the fluid contained in the pressure equipment; the chemical and physical properties necessary for operational safety must not be significantly affected within the scheduled lifetime of the equipment;
  (c) not be significantly affected by ageing;
  (d) be suitable for the intended processing procedures;
  (e) be selected in order to avoid significant undesirable effects when the various materials are put together.

4.2 (a) The pressure equipment manufacturer must define in an appropriate manner the values necessary for the design calculations referred to in 2.2.3 and the essential characteristics of the materials and their treatment referred to in 4.1;
  (b) the manufacturer must provide in his technical documentation elements relating to compliance with the materials specifications of the Directive in one of the following forms:
    – by using materials which comply with harmonized standards,
    – by using materials covered by a European approval of pressure equipment materials in accordance with Article 11,
    – by a particular material appraisal;
  (c) for pressure equipment in categories III and IV, particular appraisal as referred to in the third indent of (b) must be performed by the notified body in charge of conformity assessment procedures for the pressure equipment.

4.3 The equipment manufacturer must take appropriate measures to ensure that the material used conforms with the required specification. In particular, documentation prepared by the material manufacturer affirming compliance with a specification must be obtained for all materials.

For the main pressure-bearing parts of equipment in categories II, III, and IV, this must take the form of a certification of specific product control.

Where a material manufacturer has an appropriate quality-assurance system, certified by a competent body established with the Community and having undergone a specific assessment for materials, certificates issued by the manufacturer are presumed to certify conformity with the relevant requirement of this section.

Specific pressure equipment requirements

In addition to the applicable requirements of sections 1 to 4, the following requirements apply to the pressure equipment covered by Sections 5 and 6

5 Fired or otherwise heated pressure equipment with a risk of overheating as referred to in Article 3(1)

This pressure equipment includes:
- steam and hot-water generators as referred to in Article 3, Section 1.2, such as fired steam and hot-water boilers, superheaters and reheaters, waste-heat boilers, waste incineration boilers, electrode or immersion-type electrically heated boilers, pressure cookers, together with their accessories and where applicable their systems for treatment of feedwater and for fuel supply; and
- process-heating equipment for other than steam and hot water generation falling under Article 3, Section 1.1, such as heaters for chemical and other similar processes and pressurized food-processing equipment.

This pressure equipment must be calculated, designed and constructed so as to avoid or minimize risks of a significant loss of containment from overheating. In particular it must be ensured, where applicable that:
(a) appropriate means of protection are provided to restrict operating parameters such as heat input, heat take-off and, where applicable, fluid level so as to avoid any risk of local and general overheating,
(b) sampling points are provided where required to allow evaluation of the properties of the fluid so as to avoid risks related to deposits and/or corrosion,
(c) adequate provisions are made to eliminate risks of damage from deposits,
(d) means of safe removal of residual heat after shutdown are provided,
(e) steps are taken to avoid a dangerous accumulation of ignitable mixtures of combustible substances and air, or flame blowback.

6 Piping as referred to in Article 3, Section 1.3

Design and construction must ensure:

(a) that the risk of overstressing from inadmissible free movement or excessive forces being produced, e.g. on flanges, connections, bellows or hoses, is adequately controlled by means such as support, constraint, anchoring, alignment and pre-tension;

(b) that where there is a possibility of condensation occurring inside pipes for gaseous fluids, means are provided for drainage and removal of deposits from low areas to avoid damage from water hammer or corrosion;

(c) that due consideration is given to the potential damage from turbulence and formation of vortices; the relevant parts of 2.7 are applicable;

(d) that due consideration is given to the risk of fatigue due to vibrations in pipes;

(e) that, where fluids of Group 1 are contained in the piping, appropriate means are provided to isolate "take-off" pipes the size of which represents a significant risk;

(f) that the risk of inadvertent discharge is minimized; the take-off points must be clearly marked on the permanent side, indicating the fluid contained;

(g) that the position and route of underground piping is at least recorded in the technical documentation to facilitate safe maintenance, inspection or repair.

7 Specific quantitative requirements for certain pressure equipment

The following provisions apply as a general rule. However, where they are not applied, including in cases where materials are not specifically referred to and no harmonized standards are applied, the manufacturer must demonstrate that appropriate measures have been taken to achieve an equivalent overall level of safety.

This section is an integral part of Annex 1. The provisions laid down in this section supplement the essential requirements of Sections 1–6 for the pressure equipment to which they apply.

7.1 Allowable stresses

7.1.1 Symbols

$R_{e/t}$, yield limit, indicates the value at the calculation temperature of:
– the upper flow limit for a material presenting upper and lower flow limits;

- the 1.0% proof strength of austenitic steel and non-alloyed aluminium;
- the 0.2% proof strength in other cases.

$R_{m/20}$ indicates the minimum value of the ultimate strength 20 °C.

$R_{m/t}$ designates the ultimate strength at the calculation temperature.

7.1.2 The permissible general membrane stress for predominantly static loads and for temperatures outside the range in which creep is significant must not exceed the smaller of the following values, according to the material used:
- in case of ferritic steel including normalized (normalized rolled) steel and excluding fine-grained steel and specially heat-treated steel, 2/3 of $R_{e/t}$ and 5/12 of $R_{m/20}$;
- in the case of austenitic steel:
  - if its elongation after rupture exceeds 30%, 2/3 of $R_{e/t}$
  - or, alternatively, if its elongation after rupture exceeds 35%, 5/6 of $R_{e/t}$ and 1/3 of $R_{m/t}$;
- in the case of non-alloy or low-alloy cast steel, 10/19 of $R_{e/t}$ and 1/3 of $R_{m/20}$;
- in the case of aluminium, 2/3 of $R_{e/t}$;
- in the case of aluminium alloys excluding precipitation hardening alloys 2/3 of $R_{e/t}$ and 5/12 of $R_{m/20}$.

7.2 *Joint coefficients*

For welded joints, the joint coefficient must not exceed the following values:
- for equipment subject to destructive and non-destructive tests which confirm that the whole series of joints show no significant defects: 1;
- for equipment subject to random non-destructive testing: 0.85;
- for equipment not subject to non-destructive testing other than visual inspection: 0.7.

If necessary, the type of stress and the mechanical and technological properties of the joint must also be taken into account.

7.3 *Pressure limiting devices, particularly for pressure vessels*

The momentary pressure surge referred to in 2.11.2 must be kept to 10% of the maximum allowable pressure.

7.4 *Hydrostatic test pressure*

For pressure vessels, the hydrostatic test pressure referred to in 3.2.2 must be no less than:
- that corresponding to the maximum loading to which the pressure equipment may be subject in service taking into account its maximum allowable pressure and its maximum allowable temperature, multiplied by the coefficient 1.25, or
- the maximum allowable pressure multiplied by the coefficient 1.43, whichever is the greater.

7.5 *Material characteristics*

Unless other values are required in accordance with other criteria that must be taken into account, a steel is considered as sufficiently ductile to satisfy 4.1 (a) if, in the tensile test carried out by a standard procedure, its elongation after rupture is no less than 14% and its bending rupture energy measure on an ISO V test-piece is no less than 27 J, at a temperature not greater than 20 °C but not higher than the lowest scheduled operating temperature.

11.8.7 Declaration of conformity

In addition to ensuring that the equipment is capable of meeting the essential safety requirements of the PED, manufacturers must also complete a specified declaration of conformity (see Table 11.6) and maintain a technical file of information about how the equipment was designed and manufactured.

Table 11.6 Contents of The Declaration of Conformity

The EC declaration of conformity must contain the following particulars:
- Name and address of the manufacturer.
- Description of the pressure equipment or assembly.
- The conformity assessment procedure followed.
- For assemblies, a description of the pressure equipment constituting the assembly, and the conformity assessment procedures followed.
- Name and address of the notified body which carried out the inspection.
- Reference to the EC type examination certificate, EC design examination certificate, or EC certificate of conformity.
- Name and address of the notified body monitoring the manufacturer's quality assurance system.
- The references of the harmonized standards applied or other technical standards and specifications used.
- The references of the other Community Directives applied.
- Details of any signatory authorized to sign the legally binding declaration for the manufacturer.

11.8.8 Pressure equipment marking

Under the PED requirements, pressure equipment must be marked with at least: (see also Fig. 5.2)
- identification of the manufacturer;
- unique identification of model and serial number;
- the year of manufacture;
- maximum/minimum allowable pressure and temperature limits;
- the CE logo (see Fig. 11.2).

11.9 Pressure Equipment Regulations 1999

The Pressure Equipment Regulations (Statutory Instrument 1999 No. 2001) are the UK mechanism for implementing the PED. Section 11.9.1 shows how the Regulations are structured; the full text of all the sections is available from: http://www.hmso.gov.uk/si/si1999/19992001.htm. The authoritative version is the Queen's Printer copy published by The Stationery Office Limited as the *The Pressure Equipment Regulations 1999*, ISBN 0 11 082790 2.

11.9.1 The Pressure Equipment Regulations – structure

- All of these sections are available on
  http://www.hmso.gov.uk/si/si1999/19992001.htm

Part I – Preliminaries

1. Citation and commencement.
2. Interpretation.

Part II – Application

3. Pressure equipment and assemblies.
4. Excluded pressure equipment and assemblies.
5. Pressure equipment and assemblies placed on the market before 29 November 1999.
6. Exclusion until 30 May 2002 of pressure equipment and assemblies complying with provisions in force on 28 November 1999.

Part III – General requirements

7. General duty relating to the placing on the market or putting into service of pressure equipment.
8. General duty relating to the placing on the market or putting into service of assemblies.
9. Requirement for pressure equipment or assemblies to comply with sound engineering practice.

10 General duty relating to the supply of pressure equipment or assemblies.
11 Exceptions to placing on the market or supply in respect of certain pressure equipment and assemblies.
12 Classification of pressure equipment.
13 Conformity assessment procedures for pressure equipment.
14 Conformity assessment procedure for assemblies.
15 Exclusion for pressure equipment and assemblies for use for experimentation.
16 CE marking.
17 European approval for materials.
18 Notified bodies.
19 Recognized third-party organizations.
20 Notified bodies and recognized third-party organizations appointed by the Secretary of State.
21 Fees.
22 User inspectorates.
23 Conditions for pressure equipment and assemblies being taken to conform with the provisions of these Regulations.

Part IV – Enforcement

24 Application of Schedule 8.
25 Offences.
26 Penalties.
27 Defence of due diligence.
28 Liability of persons other than the principal offender.
29 Consequential amendments.

Schedules

Note how these schedules accurately reflect the requirements of the PED.

Schedule 1. Excluded pressure equipment and assemblies.
Schedule 2. Essential safety requirements (Annex I to the Pressure Equipment Directive).
Schedule 3. Conformity assessment tables (Annex II to the Pressure Equipment Directive).
Schedule 4. Conformity assessment procedures (Annex III to the Pressure Equipment Directive).
Schedule 5. CE marking (Annex VI to the Pressure Equipment Directive).
Schedule 6. EC declaration of conformity (Annex VII to the Pressure Equipment Directive).
Schedule 7. European approval for materials.
Schedule 8. Enforcement.

11.10 Notified Bodies

11.10.1 What are they?

Notified Bodies (NBs) are organizations designated by the national governments of the EU Member States as being competent to make independent judgments about whether or not a product complies with the essential safety requirements laid down by each CE marking directive. In order to be notified, the management structure of the organization must fulfil certain conditions and the name of the NB, along with the details of the scope of its notified activities, must be given ('notified') to the European Commission.

These NBs are also the organizations that will be requested to give judgment on the safety of a product by the enforcement authorities in the EU. They are, therefore, responsible for judging which standards should have been applied to a product and if they have been applied correctly.

Until recently, Member States could only notify bodies within their own territories and so the only bodies which were notified outside of the EEA were subsidiaries of EU resident organisations. However, under planned Mutual Recognition Agreements ('MRAs') between the EU and America, Canada, Australia, and New Zealand, regulatory authorities in these countries will be able to designate Notified Bodies within their own territory. However, for the PED, no MRAs are yet possible since there is currently no equivalent legislation to the PED or Machinery Directive in the USA or Canada. This situation may change as the 2002 implementation date draws closer.

11.10.2 UK Notified Bodies

Table 11.7 shows a sample of UK Notified Bodies for the PED. A wider listing of all European Notified Bodies is given in Appendix 2 and is available from: www.conformance.co.uk/CE_MARKING/ce_notified.html

Pressure Equipment: Directives and Legislation 315

Table 11.7 Current UK Notified Bodies

Organization	Scope of appointment	Address
APAVE UK Ltd	• Modules A1,B,B1,C1,D,D1,E,E1,H,H1 • Approval of joining procedures and staff for categories II, III, and IV • Specific designation for the issue of European approval of materials	Gothic House Barker Gate Nottingham NG1 1U5 Tel: 0115 9551880 Fax: 0115 9951881 www.apave-uk.com
B.Inst.NDT	• NDT staff approval (categories III and IV)	1 Spencer Parade Northampton NN1 5AA Tel: 01604 259056 Fax: 01604 231489 www.bindt.org
British Standards Institution	• Modules A1,B,B1,C1,D,D1,E,E1,H,H1 • Approval of joining procedures and staff for categories II, III, and IV	Marylands Avenue Hemel Hempstead Herts HP2 4SQ Tel: 01442 230442 Fax: 01442 231442 www.bsi.org.uk/services/testing/powered-equip.html
HSB Inspection Quality Ltd	• Modules A1,B,B1,C1,D,D1,E,E1,H,H1 • Approval of joining procedures and staff for categories II, III, and IV	11 Seymour Court Tudor Road Manor Park Runcorn Cheshire WA7 1SY Tel: 01928 579595 Fax: 01928 579623 www.hsbiql.co.uk
Lloyd's Register	• Modules A1,B,B1,C1,D,D1,E,E1,H,H1 • Approval of joining procedures and staff for categories II, III, and IV • Specific designation for the issue of European approval of materials	71 Fenchurch St London C3M 4BS or Hiramford Middlemarch Village Siskin Drive Coventry CV3 4FJ Tel: 0207 7099166 Fax: 0207 488 4796 or Tel: 024 76882311 Fax: 024 76305533 www.lrqa.com

Table 11.7 Cont.

Plant Safety Ltd	• Modules A1,B,B1,C1,D,D1,E,E1,H,H1 • Approval of joining procedures and staff for categories II, III, and IV • Specific designation for the issue of European approval of materials	Parklands Wilmslow Road Didsbury Manchester M20 2RE Tel: 0161 4464600 Fax: 0161 4462506 www.plantsafety.co.uk
Royal Sun Alliance Certification Services Ltd	• Modules A1,B,B1,C1,D,D1,E,E1,H,H1 • Approval of joining procedures and staff for categories II, III, and IV • Specific designation for the issue of European approval of materials	17 York St Manchester M2 3RS Tel: 0161 2353375 Fax: 0161 2353702 www.royal-and -sunalliance.com
SGS(UK) Ltd	• Modules A1,B,B1,C1,D,D1,E,E1,H,H1 • Approval of joining procedures and staff for categories II, III, and IV • Specific designation for the issue of European approval	SGS House Johns Lane Tividale Oldbury West Midlands B69 3HX Tel: 0121 5206454 Fax: 0121 5223532 www.sgs.co.uk
TUV(UK) Ltd	• Modules A1,B,B1,C1,D,D1,E,E1,H,H1 • Approval of joining procedures and staff for categories II, III, and IV • Specific designation for the issue of European approval of materials	Surrey House Surrey St Croydon CR9 1XZ Tel: 0208 6807711 Fax: 0208 6804035 www.tuv-uk.com
TWI Certification Ltd	• Approval of joining procedures and staff for categories II, III, and IV	Granta Park Great Abington Cambridge CB1 6AL Tel: 01223 891162 Fax: 01223 894219 www.twi.co.uk
Zurich Engineering Ltd	• Modules A1,B,B1,C1,D,D1,E,E1,H,H1 • Approval of joining procedures and staff for categories II, III, and IV • Specific designation for the issue of European approval of materials	54 Hagley Rd Edgbaston Birmingham B16 8QP Tel: 0121 4561311 Fax: 0121 4561754 www.zuricheng.co.uk

11.11 Sources of information

When dealing with Directives, Statutory Instruments, Regulations, and similar documents, it is important that full, up-to-date information is obtained. In many cases, the definitive content is obtainable only from the full text of the Directive, Regulation, etc. Care must, therefore, be taken to obtain these from the authorized sources. The list below shows some accepted sources of reference for pressure equipment-related topics:

11.11.1 Pressure system safety – general

Professional advice is available from the following:
- The local HSE office – the number can be obtained from directory enquiries or the phone book under *Health and Safety Executive*.

The HSE publish various books and leaflets, including:
- *Safety of pressure systems; Pressure Systems and Transportable Gas Containers Regulations 1989 Approved Code of Practice COP37* HSE Books 1990 ISBN 0 11 885514 X;
- *Guide to the Pressure Systems and Transportable Gas Container Regulations 1989* HSR30 HSE Books 1990 ISBN 0 7176 0489 6;
- *Written schemes of examination* INDG178 HSE Books 1994.

HSE priced and free publications are available by mail order from:
HSE Books, PO Box 1999, Sudbury, Suffolk CO10 6FS.
Tel: 01787 881165 Fax: 01787 313995

HSE priced publications are also available from some bookshops. For other enquiries use HSE's InfoLine Tel: 0541 545500, or write to HSE's Information Centre, Broad Lane, Sheffield S3 7HQ.

The entry website for the HSE is: http://www.hse.gov.uk/hsehome.htm

Statutory Instrument (SI) documents are available from Her Majesty's Stationery Office (HMSO) at:
www.hmso.gov.uk/legis.htm
www.hmso.gov.uk/si
www.legislation.hmso.gov.uk

There is an excellent introduction to the various Directives (and guidelines on their interpretation relating to Pressure Equipment) at:
www.tukes.fi/English/pressure/directives_and_guidelines/index.htm

11.11.2 Transportable pressure receptacles (gas cylinders)

The HSE web page for TPRs is:
www.hse.gove.uk/spd/spdtpr.htm

An important HSE press release relating to gas cylinder safety is at:
www.healthandsafety.co.uk/E00800.html

Carriage of dangerous goods (CDR) Regulations 1998 are available on:
www.hmso.gov.uk/si/si1998/19982885.htm

11.11.3 The simple pressure vessel directive/regulations

A comprehensive DTI Guidance booklet on SPV Directives/Regulations (ref URN 95/779) is available from the DTI, contact:
Peter Rutter, STRD5, Department of Trade and Industry, Room 326, 151 Buckingham Palace Road, London SW1W 9SS
Tel: 020 7215 1437
The entry web site is:
www.dti.gov.uk/strd
All DTI publications can be obtained from the DTI Publications Orderline:
Tel: 0870 15025500, Fax: 0870 1502333

11.11.4 The pressure equipment directive

Background information on mutual recognition agreements relating to European directives in general is available on
  http://europa.eu.int/comm/enterprise/international/indexb1/htm
A good general introduction to the PED is available on
  http://www.ped.eurodyne.com/directive/directive.html and
  www.dti.gov.uk/strd/pressure.htm
The European Commission Pressure Equipment Directive website has more detailed information on:
  http://europa.eu.int/comm/dg03/directs/dg3d/d2/presves/preseq.htm
  or
  http://europa.eu.int/comm/dg03/directs/dg3d/d2/presves/preseq1.htm

The CEN website also provides details of all harmonized standards:
  www.newapproach.org/directivelist.asp

The UK DTI publish a number of useful guides on the PED and these are available for download on:
  www.dti.gov.uk/strd/strdpubs.htm

A full listing of European Notified Bodies for the PED is available from:
  www.conformance.co.uk/CE_MARKING/ce_notified.html

11.11.5 The pressure equipment regulations

All the sections of these regulations are available on
http://www.hmso.gov.uk/si/si1999/19992001.htm

11.11.6 PSSRs and written schemes

The PSSRs are available on www.hmso.gov.uk/si/si2000/20000128.htm

Useful information about the ongoing requirements for pressure equipment after installation (e.g. regular inspections etc.) may also be found on the Health and Safety Executive web site at:
www.open.gov.uk/hse/hsehome.htm

The authoritative version of the Pressure Systems Safety Regulations (PSSRs) is the copy published by The Stationery Office Limited as the *The Pressure Systems Safety Regulations 2000*, ISBN 0 11 085836 0.

The United Kingdom Accreditation Service (UKAS) can advise on competent persons.

They can be contacted at:
Queens Road, Teddington Middlesex TW11 0NA.
Tel: 0181 943 7066 Fax 0181 943 7096.

The following publications give detailed information:
A guide to the Pressure Systems and Transportable Gas Containers Regulations 1989 HS(R)30 1990 HSE Books ISBN 0 11 885516 6.

Approved Code of Practice *Safety of Pressure Systems* COP 37 HSE Books 1990 ISBN 0 11 885514 X.

Approved Code of Practice Safety of Transportable Gas Containers COP 38 HSE Books 1990 ISBN 0 11 885515 8.

Introducing Competent Persons IND(S)29(L) ISBN 0 7176 0820 4 1992: HSE leaflet.

Safe Pressure Systems IND(S)27(L) ISBN 0 7176 0821 2 1991: HSE leaflet.

The Pressure Systems and Transportable Gas Containers Regulations 1989 (SI 1989 No 2169) HMSO ISBN 0 11 098169 3.
The Approved Codes of Practice list relevant HSE Guidance Notes.

The publication: *The Safe Use of Gas Cylinders*: INDG 3081/00 C150 is available free from HSE bodies Tel 01787 881165, fax 01787 313995 or www.hse.gov.uk/hsehome.htm.

CHAPTER 12

In-service Inspection

12.1 A bit of history

The in-service inspection of pressure equipment can be viewed as a completely separate subject to the requirements related to the construction and 'putting into use' of pressure equipment covered by the PED and related directives. The inspection of pressure equipment during its working life was originally initiated by engineering insurance companies with the objective of reducing the number of accidents and, therefore, claims. Over the past 100 years, in-service inspection has continued, with various statutory requirements for the periodic inspection of pressure equipment, predominantly vessels. In addition, plant users generally considered it their duty under The Health and Safety at Work Act, and the requirements of 'good practice' to have their pressure vessels inspected at regular intervals.

Matters were formalized in 1989 with the issue of The Pressure Systems and Transportable Gas Containers (PSTGC) Regulations 1989 – the main regulations covering in-service inspection of pressure systems and equipment. These regulations have jurisdiction in the UK only and were (and still are) unrelated to any EU Directives. They required the users and owners of pressure systems to demonstrate that they knew the operating pressures of their pressure systems and that the systems were actually safe at those pressures. They also placed the onus on the plant users and owners to ensure that a suitable written scheme of examination was in force for the pressure system, rather than just individual vessels.

The 1989 Regulations have been recently superseded by The Pressure Systems Safety Regulations 2000, which now apply only to pressure systems, all reference to transportable gas containers having been removed. The technical content relating to the in-service inspection of pressure system components is essentially unchanged.

12.2 The Pressure Systems Safety Regulations (PSSRs) 2000

The essence of the PSSRs is the concept that all included equipment is covered by a Written Scheme of Examination, and then that it is inspected periodically by a 'Competent Person'. Table 12.1 shows the types of pressure system equipment that are included and excluded from coverage by the PSSRs. Table 12.2 is a summary of the key points about Written Schemes of Examination.

Table 12.1 Inclusions/exclusions from the PSSRs

Pressurized systems included in the PSSRs	*Pressurized systems excluded from the PSSRs*
• A compressed air receiver and the associated pipework where the product of the pressure times the internal capacity of the receiver is greater than 250 bar litres.	• An office hot water urn (for making tea).
• A steam sterilizing autoclave and associated pipework and protective devices.	• A machine tool hydraulic system.
• A steam boiler and associated pipework and protective devices.	• A pneumatic cylinder in a compressed air system.
• A pressure cooker.	• A hand-held tool.
• A gas loaded hydraulic accumulator.	• A combustion engine cooling system.
• A portable hot water/steam cleaning unit.	• A compressed air receiver and the associated pipework where the product of the pressure times the internal capacity of the receiver is less than 250 bar litres.
• A vapour compression refrigeration system where the installed power exceeds 25 kW.	• Any pipeline and its protective devices in which the pressure does not exceed 2 bar above atmospheric pressure.
• A narrow gauge steam locomotive.	• A portable fire extinguisher with a working pressure below 25 bar at 60 °C and having a total mass not exceeding 23 kg.

Table 12.1 Cont.

- The components of self-contained breathing apparatus sets (excluding the gas container).
- A fixed LPG storage system supplying fuel for heating in a workplace.
- A portable LPG cylinder.
- A tyre used on a vehicle.

Table 12.2 Written schemes of examination (WSE) – summary sheet

What is a WSE?

It is a document containing information about selected items of plant or equipment which form a pressure system, operate under pressure, and contain a 'relevant fluid'.

What is a 'relevant fluid'?

The term 'relevant fluid' is defined in the Regulations and covers:
- compressed or liquefied gas including air above 0.5 bar pressure;
- pressurized hot water above 110 °C;
- steam at any pressure.

What does the WSE contain?

The typical contents of a written scheme of examination would include:
- identification number of the item of plant or equipment;
- those parts of the item that are to be examined;
- the nature of the examination required, including the inspection and testing to be carried out on any protective devices;
- the preparatory work necessary to enable the item to be examined;
- the date by which the initial examination is to be completed (for newly installed systems);
- the maximum interval between one examination and the next;
- the critical parts of the system which, if modified or repaired, should be examined by a competent person before the system is used again;
- the name of the competent person certifying the written scheme of examination;
- the date of certification.

Table 12.2 Cont.

The plant items included are those which, if they fail, *'could unintentionally release pressure from the system and the resulting release of stored energy could cause injury.'*

Who decides which items of plant are included in the WSE?
- The user or the owner of the equipment.
- The written scheme of examination must be 'suitable' throughout the lifetime of the plant or equipment so it needs to be reviewed periodically and, when necessary, revised.

What is 'a competent person'?
- The definition in the Regulations is: *'competent person' means a competent individual person (other than an employee) or a competent body of persons corporate or unincorporated.*

  This is probably the best available definition.

What does the 'competent person' do ?
- Advise on the nature and frequency of examination and any special safety measures necessary to prepare the system for examination; and/or
- Draw up and certify as suitable the written scheme of examination; or
- Simply certify, as suitable, a written scheme of examination prepared by the user or owner.

Users (or owners) of pressure systems are free to select any competent person they wish, but they are obliged to take all reasonable steps to ensure that the competent person selected can actually demonstrate competence.

For further information look at:
- The HSE booklet *A Guide to the Pressure Systems and Transportable Gas Containers Regulations 1989*.
- Guidance on the selection of competent persons is given in the HSE leaflet; *Introducing competent persons* IND(S)29(L) ISBN 0 7176 0820 4 1992.
- See also Section 11.11.6 of this book for further reference sources.

CHAPTER 13

References and Information Sources

13.1 European Pressure Equipment Research Council (EPERC)

EPERC: What does it do?

The principal objectives of EPERC are to meet the needs of the pressure equipment (PE) industry and to organize work in support of the European Standards. The EU provides the secretariat and each of the member countries has a national body, to act as a link with industry and research establishments. The EPERC Secretariat is administered by Jean-Bernard Veyret, Joint Research Council, Institute for Advanced Materials, PO Box 2, 1755 ZG Petten, The Netherlands: Tel +31.22.456.5238; Fax +31.22.456.3424; e-mail: veyret@jrc.nl.

The Pressure Systems Group of the UK Institution of Mechanical Engineers provides UK representatives.

Objectives

EPERC is a European network of industries, research laboratories, universities, inspection bodies, governmental institutions, and individuals set up to foster co-operative research for the greater benefit of the European industry.

The stated objectives of EPERC are to:
- establish a European network in support of the non-nuclear pressure equipment industry and small enterprises in particular;
- establish the short- and long-term research priorities of the European pressure equipment industry;
- co-ordinate co-operative research in the domain of pressure equipment, and identify funding sources for this research;

- foster technology transfer of research results to the European industry and standardization bodies. As a consequence EPERC will help to establish a European attitude on pressure equipment safety and reliability.

EPERC: Its European role

On the European level, EPERC acts to:
- draw together European expertise;
- provide a unified representation and improved image of the European PE industry;
- facilitate effective promotional and advisory roles to the EC on behalf of the PE industry;
- identify common industrial needs;
- lead to targeted and cost-effective R&D;
- benefit industry through input to standardization activities and information transfer to small and medium industries.

EPERC membership

EPERC is organized on a national network basis. A National Representative (NR) co-ordinates a national network of members in each Member Country through national meetings and events. Within this forum, the members may formulate their national viewpoints and perspectives. These national views are transmitted to EPERC by the NR through the EPERC secretariat, which is facilitated by the European Commission Joint Research Centre in Petten, The Netherlands. Members are obliged, therefore, in their own interests, to participate in their national network.

Once a year EPERC will present itself to the General Assembly (GA). This will be the forum for individual members to receive a progress report on the previous 12 months' work, comment on the performance of EPERC, and express opinion on the future direction of EPERC. This GA is designed specifically for the individual members rather than the Steering Committee or National Representatives.

EPERC technical task forces (TTFs).

TTFs are:
- TTF1 – Fatigue design;
- TTF2 – High-strength steels for pressure equipment thickness reduction;
- TTF3 – Harmonization of inspection programming in Europe;
- TTF4 – Flanges and gaskets.

If you would like to join EPERC, please request a 'Membership Form' from Katherine Lewis. Once it has been completed, return it to (i) your national representative and (ii) the EPERC Secretary.

EPERC in the UK

The UK/EPERC has 25 members, meets twice a year, and the IMechE provides the secretariat. Members are involved in the EPERC TTFs, which are responsible for initiating the work and progressing it to completion. For further information on UK/EPERC, contact Helen Ricardo: Tel 00-44-207-(973)-1273, e-mail: h_ricardo@imeche.org.uk.

13.2 European and American associations and organizations relevant to pressure equipment activities

Acronym	Organization	Contact
AA	The Aluminum Association Inc. (USA) 900 19th St NW Washington DC 20006	Tel: 00 1 (202) 862 5100 Fax: 00 1 (202) 862 5164 www.aluminum.org
ABS	American Bureau of shipping (UK) ABS House 1 Frying Pan Alley London E1 7HR	Tel: +44 (0)20 7247 3255 Fax: +44 (0)20 7377 2453 www.eagle.org
ACEC	American Consulting Engineers Council 1015, 15th St NW #802 Washington DC 20005	Tel: 00 1 (202) 347 7474 Fax: 00 1 (202) 898 0068 www.acec.org
AEAT	AEA Technology plc (UK) Harwell Didcot Oxon OX11 OQT	Tel: +44 (0)1235 821111 Fax: +44 (0)1235 432916 www.aeat.co.uk
AIE	American Institute of Engineers 1018 Aopian Way El Sobrante CA 94803	Tel: 00 1 (510) 223 8911 Fax: 00 1 (510) 223 8911 www.members-aie.org
AISC	American Institute of Steel Construction Inc 1, E Wacker Drive Suite 3100 Chicago IL 60601-2001	Tel: 00 1 (312) 670 2400 Fax: 00 1 (312) 670 5403 www.aisc.org

AISE	Association of Iron and Steel Engineers (USA) 3, Gateway Center Suite 1900 Pittsburgh PA 15222-1004	Tel: 00 1 (412) 281 6323 Fax: 00 1 (412) 2814657 www.aise.org
AISI	American Iron and Steel Institute 1101 17th SE NW Suite 1300 Washington DC 20036	Tel: 00 1 (202) 452 7100 Fax: 00 1 (202) 463 6573 www.steel.org
ANS	American Nuclear Society 555 N. Kensington Ave La Grange Park IL 60526	Tel: 00 1 (708) 352 6611 Fax: 00 1 (708) 352 0499 www.ans.org
ANSI	American National Standards Institute 11, W. 42nd St New York NY 10036	Tel: 00 1 (212) 642 4900 Fax: 00 1 (212) 398 0023 www.ansi.org
AP	APAVE UK Ltd Gothic House Barker Gate Nottingham NG1 1JU	Tel: +44 (0)115 955 1880 Fax: +44 (0)115 955 1881 www.apave-uk.com
API	American Petroleum Institute 1220 L St NW Washington DC 20005	Tel: 00 1 (202) 682 8000 Fax: 00 1 (202) 682 8232 www.api.org
ASHRAE	American Society of Heating, Refrigeration and Air Conditioning Engineers 1791 Tullie Circle NE Atlanta GA 30329	Tel: 00 1 (404) 636 8400 Fax: 00 1 (404) 321 5478 www.ashrae.org
ASME	American Society of Mechanical Engineers 3, Park Ave New York NY 10016-5990	Tel: 00 1 (973) 882 1167 Fax: 00 1 (973) 882 1717 www.asme.org

References and Information Sources

ASNT	American Society for Non-Destructive Testing 1711 Arlington Lane Columbus OH 43228-0518	Tel: 00 1 (614) 274 6003 Fax: 00 1 (614) 274 6899 www.asnt.org
ASTM	American Society for Testing of Materials 100, Barr Harbor Drive Conshohocken PA 19428-2959	Tel: 00 1 (610) 832 9585 Fax: 00 1 (610) 832 9555 www.ansi.org
AWS	American Welding Society 550 NW Le Jeune Rd Miami FLA 33126	Tel: 00 1 (305) 443 9353 Fax: 00 1 (305) 443 7559 www.awweld.org
AWWA	American Water Works Association Inc 6666 W Quincy Ave Denver CO 80235	Tel: 00 1 (303) 794 7711 Fax: 00 1 (303) 794 3951 www.awwa.org
B.Inst.NDT	British Institute of Non Destructive Testing 1 Spencer Parade Northampton NN1 5AA	Tel: +44 (0)1604 259056 Fax: +44 (0)1604 231489 www.bindt.org
BIE	British Inspecting Engineers Chatsworth Technology Park Dunston Road Chesterfield D41 8XA	Tel: +44 (0)1246 260260 Fax: +44 (0)1246 260919 www.bie-international.com
BSI	British Standards Institution Maylands Avenue Hemel Hempstead Herts HP2 4SQ	Tel: +44 (0)1442 230442 Fax: +44 (0)1442 231442 www.bsi.org.uk
BVAMA	British Valve and Actuator Manufacturers Association The MacLaren Building 35 Dale End Birmingham B4 7LN	Tel: +44 (0)121 200 1297 Fax: +44 (0)121 200 1308 www.bvama.org.uk
CEN	European Committee for Standardisation (Belgium) 36, rue de Stassart B-1050 Brussels Belgium	Tel: 00 (32) 2 550 08 11 Fax: 00 (32) 2 550 08 19 www.cenorm.be www. newapproach.org

DNV	Det Norske Veritas (UK) Palace House 3 Cathedral Street London SE1 9DE	Tel: 00 440 (207) 6080 Fax: 00 440 (207) 6048 www.dnv.com
DTI	DTI STRD 5 (UK) Peter Rutter STRD5 Department of Trade and Industry Room 326 151 Buckingham Palace Road London SW1W 9SS	Tel: +44 (0)20 7215 1437 www.dti.gov.uk/strd
DTI	DTI Publications Orderline (UK)	Tel: +44 (0)870 15025500 Fax: +44 (0)870 1502333
EC	The Engineering Council (UK) 10 Maltravers Street London WC2R 3ER	Tel: +44 (0)20 7240 7891 Fax: +44 (0)20 7240 7517 www.engc.org.uk
EIS	Engineering Integrity Society (UK) 5 Wentworth Avenue Sheffield S11 9QX.	Tel: +44 (0)114 262 1155 Fax: +44 (0)114 262 1120 http://www.demon.co.uk/e-i-s
FCI	Fluid Controls Institute Inc (USA) PO Box 1485 Pompano Beach Florida 33061	Tel: 00 1 (216) 241 7333 Fax: 00 1 (216) 241 0105 www.fluidcontrolsinstitite.org
FMG	Factory Mutual Global (USA) Westwood Executive Center 100 Lowder Brook Drive Suite 1100 Westwood MA 02090-1190	Tel: 00 1 (781) 326 5500 Fax: 00 1 (781) 326 6632 www.fmglobal.com
HMSO	Her Majesty's Stationery Office	www.hmso.gov.uk/legis.htm www.hmso.gov.uk/si www.legislation.hmso.gov.uk
HSB	Hartford Steam Boiler (USA) PO Box 61509 King of Prussia PA 19406-0909 USA	Tel: 00 1 (484) 582 1866 Fax: 00 01 (484) 582 1802 www.hsb.com

HSE	HSE Books (UK) PO Box 1999 Sudbury Suffolk CO10 6FS	Tel: +44 (0)1787 881165 Fax: +44 (0)1787 313995 www.hse.gov.uk/hsehome.htm
HSE	HSE's InfoLine (fax enquiries) HSE Information Centre Broad Lane Sheffield S3 7HQ	Fax: +44 (0)114 2892333 www.hse.gov.uk/hsehome.htm
HTRI	Heat Transfer Research Inc (USA) 1500 Research Parkway Suite 100 College Station Texas 77845	Tel: 00 1 (409) 260 6200 Fax: 00 1 (409) 260 6249 www.htrinet.com
IMechE	The Institution of Mechanical Engineers (UK) 1 Birdcage Walk London SW1H 9JJ	Tel: +44 (0)20 7222 7899 Fax: +44 (0)20 7222 4557 www.imeche.org.uk
IoC	The Institute of Corrosion (UK) 4 Leck Street Leighton Buzzard Bedfordshire LU7 9TQ	Tel: +44 (0)1525 851 771 Fax: +44 (0)1525 376 690 www.icorr.demon.co.uk
IoE	The Institute of Energy (UK) 18 Devonshire Street London W1N 2AU	Tel: +44 (0)20 7580 7124 Fax: +44 (0)20 7580 4420 http://www.instenergy.org.uk
IoM	The Institute of Materials (UK) 1 Carlton House Terrace London SW1Y 5DB	Tel: +44 (0)20 7451 7300 Fax: +44 (0)20 7839 1702 http://www.instmat.co.uk
IPLantE	The Institution of Plant Engineers (UK) 77 Great Peter Street Westminster London SW1P 2EZ	Tel: +44 (0)20 7233 2855 Fax: +44 (0)20 7233 2604 http://www.iplante.org.uk
IQA	The Institute of Quality Assurance (UK) 12 Grosvenor Crescent London SW1X 7EE	Tel: +44 (0)20 7245 6722 Fax: +44 (0)20 7245 6755 http://www.iqa.org

ISO	International Standards Organization (Switzerland) PO Box 56 CH-1211 Geneva Switzerland	Tel: 00 (22) 749 011 Fax: 00 (22) 733 3430 www.iso.ch
LR	Lloyd's Register (UK) 71 Fenchurch St London EC3M 4BS	Tel: +44 (0)20 7709 9166 Fax: +44 (0)20 7488 4796 www.lrqa.com
MSS	Manufacturers Standardization Society of the Valve and Fittings Industry (USA) 127 Park Street NE Vienna VA 22180-4602	Tel: 00 1 (703) 281 6613 Fax: 00 1 (703) 281 6671 www.mss-hq.com
NACE	National Association of Corrosion Engineers (USA) 1440 South Creek Drive Houston TX 7708	Tel: 00 1 (281) 228 6200 Fax: 00 1 (281) 228 6300 www.nace.org
NBBPVI	National Board of Boiler and Pressure Vessel Inspectors (USA) 1155 North High Street Columbus Ohio 43201	Tel: 00 1 (614) 888 8320 Fax: 00 1 (614) 888 0750 www.nationalboard.org
NFP	National Fire Protection Association (USA) 1, Batterymarch Park PO Box 9101 Quincy MA 02269-9101	Tel: 00 1 (617) 770 3000 Fax: 00 1 (617) 770 0700 www.nfpa.org
NFU	National Fluid Power Association (USA) 3333 N Mayfair Rd Milwaukee WI 53222-3219	Tel: 00 1 (414) 778 3344 Fax: 00 1 (414) 778 3361 www.nfpa.com
NIST	National Institute of Standards and Technology (USA) 100 Bureau Drive Gaithersburg MD 20899-0001	Tel: 00 1 (301) 975 8205 Fax: 00 1 (301) 926 1630 www.nist.gov

PFI	Pipe Fabrication Institute (USA) 655 - 32nd Ave Suite 201 Lachine Qc. Canada HT8 3G6	Tel: 00 1 (514) 634 3434 Fax: 00 1 (514) 634 9736 www.pfi-institute.org
PS	Plant Safety Ltd (UK) Parklands Wilmslow Road Didsbury Manchester M20 2RE	Tel: +44 (0)1614 464600 Fax: +44 (0)1614 462506 www.plantsafety.co.uk
RAB	Registrar Accreditation Board (USA) PO Box 3003 Milwaukee WI 53201-3005	Tel: 00-1- (888)-722-2440 Fax: 00-1- (414)-765-8661 www.rabnet.com
RSA	Royal Sun Alliance Certification Services Ltd (UK) 17 York St Manchester M2 3RS	Tel: +44 (0)161 235 3375 Fax: +44 (0)161 235 3702 www.royal-and-sunalliance.com/
SAFeD	Safety Assessment Federation (UK) Nutmeg House 60 Gainsford Street Butlers Wharf London SE1 2NY	Tel: +44 (0)20 7403 0987 Fax: +44 (0)20 7403 0137 www.safed.co.uk
SGS	SGS (UK) Ltd SGS House Johns Lane Tividale, Oldbury W Midlands B69 3HX	Tel: +44 (0)121 5206 454 Fax: +44 (0)121 5223 532 www.sgs.com
TEMA	Tubular Exchanger Manufacturers Association Inc 25 N Broadway Tarrytown New York 10591	Tel: 00 1 (914) 332 0040 Fax: 00 1 (914) 332 1541 www.tema.org

TUV	TUV (UK) Ltd Surrey House Surrey St Croydon CR9 1XZ	Tel: +44 (0)20 8680 7711 Fax: +44 (0)20 8680 4035 www.tuv-uk.com
TWI	TWI Certification Ltd (UK) Granta Park Great Abington Cambridge CB1 6AL	Tel: +44 (0)1223 891162 Fax: +44 (0)1223 894219 www.twi.co.uk
UKAS	The United Kingdom Accreditation Service 21-47 High Street Feltham Middlesex TW13 4UN	Tel: +44 (0)20 8917 8554 Fax: +44 (0)20 8917 8500 www.ukas.com
WJS	The Welding and Joining Society (UK) Granta Park Great Abington Cambridge CB1 6AL	Tel: +44 (0)1223 891162 Fax: +44 (0)1223 894219 http://www.twi.co.uk/ members.wjsinfo.html
ZURICH	Zurich Engineering Ltd (UK) 54 Hagley Rd Edgbaston Birmingham B16 8QP	Tel: +44 (0)121 4561311 Fax: +44 (0)121 4561754 www.zuricheng.co.uk

13.3 Pressure vessel technology references

1. Cross, W. (1990) *The Code, An Authorized History of the ASME Boiler and Pressure Vessel Code*, ASME, New York.
2. Bernstein, M. D. (1998) Design criteria for boilers and pressure vessels in the USA, *J. Pressure Vessel Technol.*, **110**, 430–443.
3. Moen, R. A. (1996) *Practical Guide to ASME Section II – 1996 Materials Index*, Casti Publishing Inc., Edmonton, Alberta, Canada.
4. *Metals and Alloys in the Unified Numbering System*, Sixth Edition (1993), Society of Automotive Engineers, Warrendale, PA.
5. *Pressure Relief Devices* (1994) ASME Performance Test Code PTC 25.
6. Bernstein, M. D. and Friend, R. G. (1995) ASME Code safety valve rules – a review and discussion, *J. Pressure Vessel Technol.*, **117**, 104–114.
7. Narayanan, T. V. (1993) Criteria for approving equipment for continued operation, welding research council bulletin 380, Recommendations to ASME for Code guidelines and criteria for continued operation of equipment, Welding Research Council, New York, pp. 9–25.
8. Harth, G. and Sherlock, T. (1985) Monitoring the service induced damage in utility boiler pressure vessels and piping systems, *Proceedings 1985 PVP Conference*, Vol. 98-1, ASME, New York.
9. Megyesy, E. F. and Buthod P. (1998) *Pressure Vessel Handbook*, Eleventh Edition, Pressure Vessel Handbook Publishers.
10. Krishna, P. S. (1984) *Mechanical Design of Heat Exchangers and Pressure Vessel Components*, Acturus Publishers.
11. Spence, J. and Tooth, A. S. (1994) *Pressure Vessel Design: Concepts and Principles*, E & FN Spon.
12. Bednar, H. H. (1991) *Pressure Vessel Design Handbook*, Second Edition, Krieger Publishing Company.
13. Moss, D. R. (1997) *Pressure Vessel Design Manual: Illustrated Procedures for Solving Major Pressure Vessel Design Problems*, Second Edition, Gulf Publishing Corporation.
14. Steingress, F. M. and Frost, H. J. (1994) *High-pressure Boilers*, Second Edition, American Technical Publishers.
15. Steingress, F. M. (1994) *Low-pressure Boilers*, Third Edition, American Technical Publishers.

16. Bernstein, M. D., Yoder, L. W. (1998) *Power Boilers: A Guide to Section I of the ASME Boiler and Pressure Vessel Code*, ASME, New York.
17. Cross, W. and Pensky A. J. (1990) *The Code: An Authorized History of the ASME Boiler and Pressure Vessel Code*, ASME, New York.
18. Port, R. D. and Herro, H. M. (1991) *The Nalco Guide to Boiler Failure Analysis*, McGraw Hill Text.
19. Inspection of Fired Boilers and Heaters *(Rp 573)* (1991), American Petroleum Institute.
20. International Design Criteria of Boilers and Pressure Vessels (1984) AMSE, New York.

Appendix 1

Steam Properties Data

Pressure [kPa]	Temp. [°C]	Specific volume [m^3/kg] Sat. liquid [M^3/kg]	vfg [M^3/kg]	Sat. Vapor [M^3/kg]	Internal Energy [kJ/kg] uf [kJ/kg]	ug [kJ/kg]	ufg [kJ/kg]	Enthalpy [kJ/kg] hf [kJ/kg]	hfg [kJ/kg]	hg [kJ/kg]
0.6113	0.01	0.00109	206.09891	206.1	0	2375.011736	2375.011736	0.000666317	2501	2501.000666
0.813	4	0.001	157.229	157.23	16.77	2379.942823	2363.172823	16.770813	2491	2507.770813
0.8721	5	0.0010001	147.1189999	147.12	20.97	2382.66752	2361.69752	20.97087219	2490	2510.970872
0.935	6	0.0010001	137.7289999	137.73	25.19	2383.413385	2358.223385	25.19093509	2487	2512.190935
1.072	8	0.0010002	120.9189998	120.92	33.59	2385.964832	2352.374832	33.59107221	2482	2515.591072
1.2276	10	0.0010004	106.3789996	106.38	42	2389.10914	2347.10914	42.00122809	2477.7	2519.701228
1.312	11	0.0010004	99.8559996	99.857	46.2	2390.588929	2344.388929	46.20131252	2475.4	2521.601313
1.402	12	0.0010005	93.7829995	93.784	50.41	2391.926235	2341.516235	50.4114027	2473	2523.411403
1.497	13	0.0010007	88.1229993	88.124	54.6	2393.37987	2338.77987	54.60149805	2470.7	2525.301498
1.598	14	0.0010008	82.8469992	82.848	59.79	2395.700495	2335.910495	59.79159928	2468.3	2528.091599
1.7051	15	0.0010009	77.9249991	77.926	62.99	2396.020084	2333.030084	62.99170663	2465.9	2528.891707
1.818	16	0.0010011	73.3319989	73.333	67.19	2397.472426	2330.282426	67.19182	2463.6	2530.79182
1.938	17	0.0010012	69.0429988	69.044	71.38	2398.774668	2327.394668	71.38194033	2461.2	2532.58194
2.064	18	0.0010014	65.0369986	65.038	75.57	2400.133635	2324.563635	75.57206689	2458.8	2534.372067
2.198	19	0.0010016	61.2919984	61.293	79.76	2401.540188	2321.780188	79.76220152	2456.5	2536.262202
2.339	20	0.0010018	57.7899982	57.791	83.95	2402.879194	2318.929194	83.95234321	2454.1	2538.052343
2.487	21	0.001002	54.512998	54.514	88.14	2404.366174	2316.226174	88.14249197	2451.8	2539.942492

Steam Properties Data

2.645	22	0.0010022	51.4459978	51.447	92.32	2313.325336	2405.645336	92.32265082	2449.4	2541.722651
2.81	23	0.0010024	48.5729976	48.574	96.51	2310.509877	2407.019877	96.51281674	2447	2543.512817
2.985	24	0.0010027	45.8819973	45.883	100.7	2307.742238	2408.442238	100.7029931	2444.7	2545.402993
3.169	25	0.0010029	43.3589971	43.36	104.88	2304.895338	2409.775338	104.8831782	2442.3	2547.183178
3.363	26	0.0010032	40.9929968	40.994	109.06	2302.040552	2411.100552	109.0633738	2439.9	2548.963374
3.567	27	0.0010035	38.7729965	38.774	113.25	2299.296721	2412.546721	113.2535795	2437.6	2550.853579
3.782	28	0.0010037	36.6889963	36.69	117.42	2296.442216	2413.862216	117.423796	2435.2	2552.623796
4.008	29	0.001004	34.731996	34.733	121.6	2293.59416	2415.19416	121.604024	2432.8	2554.404024
4.246	30	0.0010043	32.8929957	32.894	125.78	2290.83634	2416.61634	125.7842643	2430.5	2556.284264
4.496	31	0.0010046	31.1639954	31.165	129.96	2287.986677	2417.946677	129.9645167	2428.1	2558.064517
4.759	32	0.001005	29.538995	29.54	134.14	2285.123923	2419.263923	134.1447828	2425.7	2559.844783
5.034	33	0.0010053	28.0099947	28.011	138.32	2282.397687	2420.717687	138.3250607	2423.4	2561.725061
5.324	34	0.0010056	26.5699944	26.571	142.5	2268.54135	2411.04135	142.5053538	2410	2552.505354
5.628	35	0.001006	25.214994	25.216	146.67	2276.690014	2423.360014	146.6756618	2418.6	2565.275662
5.947	36	0.0010063	23.9389937	23.94	150.85	2273.834804	2424.684804	150.8559845	2416.2	2567.055984
6.632	38	0.0010071	21.6009929	21.602	159.2	2268.242215	2427.442215	159.2066791	2411.5	2570.706679
7.384	40	0.0010078	19.5219922	19.523	167.56	2262.54961	2430.10961	167.5674416	2406.7	2574.267442
9.593	45	0.0010099	15.2569901	15.258	188.44	2248.439694	2436.879694	188.449688	2394.8	2583.249688
12.35	50	0.0010121	12.0309879	12.032	209.32	2234.117299	2443.437299	209.3324994	2382.7	2592.032499
15.78	55	0.0010146	9.5669854	9.568	230.21	2219.73297	2449.94297	230.2260104	2370.7	2600.92601
19.94	60	0.0010172	7.7669828	7.671	251.11	2205.560543	2456.670543	251.130283	2358.5	2609.630283
25.03	65	0.0010199	6.1959801	6.197	272.02	2191.114618	2463.134618	272.0455281	2346.2	2618.245528

31.19	70	0.0010228	5.0409772	5.042	292.95	2176.571921	2469.521921	292.9819011	2333.8	2626.781901
38.58	75	0.0010259	4.1299741	4.131	313.9	2162.065599	2475.965599	313.9395792	2321.4	2635.339579
47.39	80	0.0010291	3.4059709	3.407	334.86	2147.391039	2482.251039	334.908769	2308.8	2643.708769
57.83	85	0.0010325	2.8269675	2.828	355.84	2132.516469	2488.356469	355.8997095	2296	2651.899709
70.14	90	0.001036	2.359964	2.361	376.85	2117.672125	2494.522125	376.922665	2283.2	2660.122665
84.55	95	0.0010397	1.9809603	1.982	397.88	2102.709807	2500.589807	397.9679066	2270.2	2668.167907
101.35	100	0.0010435	1.6719565	1.673	418.94	2087.547209	2506.487209	419.0457587	2257	2676.045759
143.27	110	0.0010516	1.2089484	1.21	461.14	2056.993963	2518.133963	461.2906627	2230.2	2691.490663
198.53	120	0.0010603	0.8908397	0.8919	503.5	2025.741594	2529.241594	503.7105014	2202.6	2706.310501
270.1	130	0.0010697	0.6674303	0.6685	546.02	1993.927076	2539.947076	546.308926	2174.2	2720.508926
361.3	140	0.0010797	0.5078203	0.5089	588.74	1961.224526	2549.964526	589.1300956	2144.7	2733.830096
475.8	150	0.0010905	0.3917095	0.3928	631.68	1927.92462	2559.60462	632.1988599	2114.3	2746.49886
617.8	160	0.001102	0.305998	0.3071	674.86	1893.554436	2568.414436	675.5408156	2082.6	2758.140816
791.7	170	0.0011143	0.2416857	0.2428	718.33	1858.157431	2576.487431	719.2121913	2049.5	2768.712191
1002.1	180	0.0011274	0.1929726	0.1941	762.09	1821.622158	2583.712158	763.2197675	2015	2778.219768
1254.4	190	0.0011414	0.1553586	0.1565	806.19	1783.918172	2590.108172	807.6217722	1978.8	2786.421772
1554	200	0.0011565	0.1262435	0.1274	850.65	1744.517601	2595.167601	852.447201	1940.7	2793.147201
1906	210	0.0011726	0.1032274	0.1044	895.53	1703.948576	2599.478576	897.7649756	1900.7	2798.464976
2318	220	0.00119	0.085	0.08619	940.87	1661.47	2602.34	943.62842	1858.5	2802.12842
2795	230	0.0012088	0.0703712	0.07158	987.74	1617.112496	2604.852496	991.118596	1813.8	2804.918596
3344	240	0.0012291	0.0585309	0.05976	1033.2	1569.77267	2602.97267	1037.31011	1765.5	2802.81011
3973	250	0.0012512	0.0488788	0.05013	1080.4	1522.004528	2602.404528	1085.371018	1716.2	2801.571018

Steam Properties Data

4688	260	0.0012755	0.0409345	0.04221	1128.4	1470.599064	2598.999064	1134.379544	1662.5	2796.879544
5499	270	0.0013023	0.0343377	0.03564	1177.4	1416.376988	2593.776988	1184.561348	1605.2	2789.761348
6412	280	0.0013321	0.0288379	0.03017	1227.5	1358.691385	2586.191385	1236.041425	1543.6	2779.641425
7436	290	0.0013656	0.0242044	0.02557	1278.9	1297.116082	2576.016082	1289.054602	1477.1	2766.154602
8581	300	0.0014036	0.0202664	0.02167	1332	1230.994022	2562.994022	1344.044292	1404.9	2748.944292
9856	310	0.001447	0.016903	0.01835	1387.7	1116.904032	2504.604032	1401.961632	1283.5	2685.461632
11274	320	0.0014988	0.0139912	0.01549	1444.6	1080.863211	2525.463211	1461.497471	1238.6	2700.097471
12845	330	0.001561	0.011435	0.012996	1505.3	993.717425	2499.017425	1525.351045	1140.6	2665.951045
14586	340	0.0016379	0.0091621	0.0108	1573	894.2616094	2467.261609	1596.890409	1027.9	2624.790409
16513	350	0.00174	0.007073	0.008813	1641	776.603551	2417.603551	1669.73262	893.4	2563.13262
18651	360	0.0018925	0.0050525	0.006945	1725.3	626.2658225	2351.565823	1760.597018	720.5	2481.097018
21030	370	0.002213	0.002712	0.004925	1848.8	384.56664	2233.36664	1895.33939	441.6	2336.93939
22090	374.14	0.003155	0	0.003155	2029.6	0	2029.6	2099.29395	0	2099.29395

APPENDIX 2

Some European Notified Bodies (PED)

Two lists of PED Institutions are presented in this appendix. These are:
1. Notified bodies (PED Article 12)
2. Recognized third-party organizations (PED Article 13).

Notified Bodies (PED Article 12)

EAM = notified also for European Approval of Materials (Article 11)
OP = notified also for approval of operating procedures of permanent joints (Annex I, 3.1.2)
PP = notified also for approval of personnel who make permanent joints (Annex I, 3.1.2)

Notified Bodies

No.	Body	State	EAM	OP	PP
0026	AIB-Vincotte International 27-29, André Drouart B-1160 Bruxelles	Belgium	x	x	x
0029	APRAGAZ a.s.b.l. Chaussée de Vilvorde 156 B-1120 Bruxelles	Belgium	x	x	x
0030	Arbejdstilsynet Landskronagade, 33 DK- København Ø	Denmark	x	x	x
0034	TÜV Saarland Saarbrücker Str. 8 D-66280 Sulzbach	Germany	x	x	x
0036	TÜV Süddeutschland Bau und Betrieb GmbH Westendstr. 199 D-80686 München	Germany	x	x	x
0037	Zurich Certification Limited 54, Hagley Road Edgbaston Birmingham B16 8QP	United Kingdom	x	x	x
0038	Lloyd's Register of Shipping 29 Wellesley Road Croydon CRO 2AJ	United Kingdom	x	x	x
0040	Royal & Sun Alliance Certification Services 17 York Street Manchester M2 3RS	United Kingdom	x	x	x
0041	Plant Safety Limited Parklands Wilmslow Road Didsbury Manchester M20 2RE	United Kingdom	x	x	x
0043	TÜV Pfalz EV Merkurstr. 45 D-67663 Kaiserslautern	Germany	x	x	x
0044	RWTÜV EV Steubenstr. 53 D-45138 Essen	Germany	x	x	x

Some European Notified Bodies (PED)

0045	TÜV Nord EV Grosse Bahnstrasse 31 D-22525 Hamburg	Germany	x	x	x
0060	GAPAVE 191, rue de Vaugirard F - 75015 Paris	France	x	x	x
0062	Bureau Veritas 17 bis, place des reflets La Defense 2 F 92400 Courbevoie	France	x	x	x
0085	DVGW Deutscher Verein des Gas- und Wasserfaches EV Josef-Wirmer-Str. 1-3 D-53123 Bonn	Germany			
0086	British Standards Institution Quality Assurance BSI Marylands Avenue Hemel Hempstead Hertfordshire HP2 4SQ	United Kingdom	x	x	x
0090	TÜV Thüringen EV Melchendorfer Str. 64 D-99096 Erfurt	Germany	x	x	x
0091	TÜV Technische Überwachung Hessen GmbH Rüdesheimer Str. 119 D - 64285 Darmstadt	Germany	x	x	x
0094	Lloyd's Register Espana, S.A. C/ Princesa, 29-1° E - 28008 Madrid	Spain	x	x	x
0098	Germanischer Lloyd AG Vorsetzen 32 D-20459 Hamburg	Germany		x	
0100	Instituto Superiore Prevenzione E Sicurezza Del Lavoro Ispesl Via Alesandria, 200 I - 00198 Roma	Italy	x	x	x
0200	Force-Dantest Cert Park Alle 345 DK - 2605 Brøndby	Denmark	x	x	x

0343	Stoomwezen B.V. DSW Weena Zuid 168 Postbus 769 NL - 3000 AT Rotterdam	Netherland	x	x	x
0408	TÜV Österreich Krugerstrasse 16 A - 1015 Wien	Austria	x	x	x
0424	Inspecta OY PL 44 FIN - 00811 Helsinki	Finland		x	x
0434	Det Norske Veritas Region Norge AS (DVN RN) DVN RN Veritasveien 1 N - 1322 Høvik	Norway	x	x	x
0472	AMT Für Arbeitsschutz Adolph-Schönfelder-Str. 5 D - 22083 Hamburg	Germany		x	x
0525	Lloyd's Register Quality Assurance GmbH Mönckebergstr. 27 D-20095 Hamburg	Germany	x	x	x
0526	CETIM 52 avenue Félix Louat BP 80067 F 60304 Senlis Cedex	France	x only		
0531	TÜV BAYERN SZA Arsenal, Objekt 207 A-1030 Wien	Austria	x	x	x
0532	TPA-Energie-Und Umwelttechnik GmbH TPA Laxenburgerstrasse, 228 A-1230 Wien	Austria	x	x	x
0671	TÜV Anlagentechnik GmbH Unternehmungsgruppe TÜV Rheinland/Berlin- Brandenburg Am Grauen Stein D-51105 Köln	Germany	x	x	x
0685	DEKRA Automobil AG Handwerkstrasse 15 D-70565 Stuttgart	Germany		x	

Some European Notified Bodies (PED)

0686	Zentrallabor der TOS E V - Zertifizierungsstelle für Aufzüge Fischerweg 408 D - 18069 Rostock	Germany	x	x	x
0851	Association pour la sécurité des appareils à pression ASAP Tour Aurore, 18 Place des reflets F - 92975 Paris la Defense 2 Cedex	France	x	x	x
0871	Hartford Steam Boiler International GmbH Frerener Str. 13 D - 49785 Lingen (Ems)	Germany		x	
0874	Signum Gesellschaft Für Anlagensicherheit GmbH Industriepark Höchst, Geb. K801 D - 65926 Frankfurt	Germany		x	
0875	Polartest OY P.O. Box 41 FIN - 01621 Vantaa	Finland		x	x
0878	HSB Inspection Quality Ltd 11 Seymour Court Tudor Park Manor Park Runcorn Cheshire WA7 1SY	United Kingdom		x	x
0879	TÜV UK Limited Surrey House Surrey Street Croydon CR9 1XZ	United Kingdom	x	x	x

Recognized third-party organizations (PED Article 13)

OP = Approval of operating procedures of permanent joints (Annex I, 3.1.2).
PP = Approval of personnel who make permanent joints (Annex I, 3.1.2).
NDT = Approval of NDT personnel (Annex I, 3.1.3).

Recognized Third-Party Organizations

Body	State	OP	PP	NDT
British Institute of Non-Destructive Testing 1 Spencer Parade Northampton NN1 5AA	United Kingdom			x
COFREND 1, rue Gaston Boissier F 75724 Paris Cedex 15	France			x
Deutsche Gesellschaft für Zerstörungsfreie Prüfung EV - DGZFP Motardstrasse 54 D-13629 Berlin	Germany			x
Deutscher Verband für Schweissen Und Verwandte Verfahren EV DVS-PERSZERT IM DVS Aachener Strasse 172 D-40223 Düsseldorf	Germany		x	
DVS Zert EV Aachener Str. 172 D-40223 Düsseldorf	Germany	x		
Inspecta OY PL 44 00811 Helsinki	Finland			x

Some European Notified Bodies (PED)

Lloyd's Register Of Shipping 29 Wellesley Road Croydon CRO 2AJ	United Kingdom			x
Lloyd's Register Quality Assurance GmbH Mönckebergstr. 27 D-20095 Hamburg	Germany	x	x	x
Österreichische Gesellschaft für Zerstörungsfreie Prüfung ÖGFZP Krugerstrasse 16 A-1015 Wien	Austria			x
Royal & Sun Alliance Certification Services 17 York Street Manchester M2 3RS	United Kingdom			x
RWTÜV EV Steubenstr. 53 D-45138 Essen	Germany	x	x	x
AF Kontroll AB Box 8133 SE-104 20 Stockholm	Sweden			x
SAQ OFP Certifiering AB Box 49306 SE-100 29 Stockholm	Sweden			x
SLV Berlin-Brandenburg GmbH Luxemburgerstrasse 21 D-13353 Berlin	Germany	x	x	
SLV Duisburg GmbH Bismarckstrasse 85 D-47057 Duisburg	Germany	x	x	
SLV Halle GmbH Köthenerstr. 33a D-06118 Halle	Germany	x	x	
SLV Hannover Am Lindener Hafen 1 D-30453 Hannover	Germany	x	x	

SLV Mannheim GmbH Käthe-Kollwitz-Str. 19 D-68169 Mannheim	Germany	x	x	
SLV Mecklenburg-Vorpommern GmbH Erich-Schlesinger-Strasse 50 D-18059 Rostock	Germany	x	x	
SLV Saarland GmbH Heuduckstrasse 91 D-66117 Saarbrücken	Germany	x	x	
SECTOR CERT GmbH Kirchstrasse 12 D-53840 Troisdorf	Germany			x
TÜV anlagentechnik GmbH Unternehmungsgruppe TÜV Rheinland/Berlin-Brandenburg Am Grauen Stein D-51105 Köln	Germany	x	x	x

APPENDIX 3

Standards and Directives Current Status

The tables in this appendix summarize the reported status of European Directives and related technical standards for the following items of pressure-related equipment:
- flanges and joints
- GRP tanks and vessels
- piping and pipelines
- valves
- shell and water-tube boilers
- steel castings
- steel forgings
- steel tubes and fittings
- pressure equipment steels – general
- structural steels
- unfired pressure vessels
- transportable gas equipment

Each standard and directive is referred to the CEN technical body responsible for its development, and the corresponding approval dates and status.

CEN/TC 74 (Flanges and their joints)

Subject	Work Item ID	Standard reference	Directive(s)	Technical body	Start approval	Ratified	Current status
Flanges and their joints – Bolting – Part 1: Selection of bolting	74028	EN 1515-1: 1999	97/23/EC	CEN/TC 74	–	25/09/1999	Ratified
Flanges and their joints – Bolting – Part 2: Combination of flange and bolting materials for steel flanges – PN designated	74029	prEN 1515-2	97/23/EC	CEN/TC 74	–	00:00:00	Under approval
Flanges and their joints – Circular flanges for pipes, valves, fittings and accessories – Class designated – Flanges with bore sizes DN 650 to DN 1500	74034	–	97/23/EC	CEN/TC 74	–	–	Under development
Flanges and their joints – Circular flanges for pipes, valves, fittings and accessories PN-designated – Part 1: Steel flanges	74005	prEN 1092-1	97/23/EC 89/106/EEC	CEN/TC 74	–	00:00:00	Under approval
Flanges and their joints – Circular flanges for pipes, valves, fittings and accessories, Class designated – Part 1: Steel flanges NPS 1/2 to NPS 24 (DN 15 to DN 600).	74006	–	97/23/EC	CEN/TC 74	–	–	Under development
Flanges and their joints – Circular flanges for pipes, valves, fittings and accessories, PN designated – Part 4: Aluminium alloy flanges	74011	prEN 1092-4	97/23/EC 89/106/EEC	CEN/TC 74	–	00:00:00	Under approval

Standards and Directives – Current Status

Description	Number	Reference	Directive	Committee		Date	Status
Flanges and their joints – Circular flanges for pipes, valves, fittings and accessories, class designated – Part 4: Aluminium alloy flanges	74033	prEN 1759-4	97/23/EC	CEN/TC 74	–	00:00:00	Under approval
Flanges and their joints – Design rules for gasketed circular flange connections – Part 1: Calculation method	74040	prEN 1591-1	97/23/EC	CEN/TC 74	–	00:00:00	Under approval
Flanges and their joints – Design rules for gasketed circular flange connections – Part 2: Gasket parameters	74041	prENV 1591-2	97/23/EC	CEN/TC 74	–	00:00:00	Under approval
Flanges and their joints – Determination of the calculation method for P/T ratings	74018	–	97/23/EC	CEN/TC 74	–	–	Under development
Flanges and their joints – Dimensions of gaskets for PN-designated flanges – Part 1: Non-metallic flat gaskets with or without inserts	74019	EN 1514-1: 1997	97/23/EC	CEN/TC 74	–	09/01/1997	Ratified
Flanges and their joints – Dimensions of gaskets for PN-designated flanges – Part 2: Spiral wound gaskets for use with steel flanges	74020	EN 1514-2: 1997	97/23/EC	CEN/TC 74	–	09/01/1997	Ratified
Flanges and their joints – Dimensions of gaskets for PN-designated flanges – Part 3: Non-metallic PTFE envelope gaskets	74021	EN 1514-3: 1997	97/23/EC	CEN/TC 74	–	09/01/1997	Ratified

Description	Number	Reference	Directive	Committee		Date	Status
Flanges and their joints – Dimensions of gaskets for PN-designated flanges – Part 4: Corrugated, flat or grooved metallic and filled metallic gaskets for use with steel flanges	74022	EN 1514-4: 1997	97/23/EC	CEN/TC 74	–	09/01/1997	Ratified
Flanges and their joints – Gasket parameters and test procedures relevant to the design rules for gasketed circular flange connections	74032	prEN 13555	97/23/EC	CEN/TC 74	–	00:00:00	Under approval
Flanges and their joints – Gasket parameters and test procedures relevant to the design rules for gasketed circular flange connections	74023	prEN 12560-1	97/23/EC	CEN/TC 74	–	00:00:00	Under approval
Flanges and their joints – Gaskets for Class-designated flanges – Part 2: Spiral wound gaskets for use with steel flanges	74024	prEN 12560-2	97/23/EC	CEN/TC 74	–	00:00:00	Under approval
Flanges and their joints – Gaskets for Class-designated flanges – Part 3: Non-metallic PTFE envelope gaskets	74025	prEN 12560-3	97/23/EC	CEN/TC 74	–	00:00:00	Under approval
Flanges and their joints – Gaskets for Class-designated flanges – Part 4: Corrugated, flat or grooved metallic and filled metallic gaskets for use with steel flanges	74026	prEN 12560-4	97/23/EC	CEN/TC 74	–	00:00:00	Under approval

Flanges and their joints – Gaskets for Class-designated flanges – Part 5: Metallic ring-joint gaskets for use with steel flanges	74027	prEN 12560-5	97/23/EC	CEN/TC 74	–	00:00:00	Under approval
Flanges and their joints – Gaskets for PN designated flanges – Part 6: Kammprofile gaskets	74036	–	97/23/EC	CEN/TC 74	–	–	Under development
Flanges and their joints – Gaskets for PN designated flanges – Part 7: Covered metal jacketed gaskets	74038	–	97/23/EC	CEN/TC 74	–	–	Under development
Flanges and their joints – Gaskets for class designated flanges – Part 6: Kammprofile gaskets	74037	–	97/23/EC	CEN/TC 74	–	–	Under development
Flanges and their joints – Gaskets for class designated flanges – Part 7: Covered metal jacketed gaskets	74039	–	97/23/EC	CEN/TC 74	–	–	Under development
Flanges and their joints – Quality assurance standards for individual gasket types for industrial applications	74031	–	97/23/EC	CEN/TC 74	–	–	Under development
Pipework components – Definition and selection of DN (nominal size) (ISO 6708:1995)	74002	EN ISO 6708: 1995	97/23/EC 89/106/EEC	CEN/TC 74	–	12/06/1995	Ratified
Pipework components – Definition and selection of PN	74003	EN 1333: 1996	97/23/EC 89/106/EEC	CEN/TC 74	–	29/03/1996	Ratified

CEN/TC 210 (GRP tanks and vessels)

Subject	Work item ID	Standard reference	Directive(s)	Technical body	Start approval	Ratified	Current status
Filament wound FRP pressure vessels – Materials, design, calculation, manufacturing and testing	210013	–	97/23/EC	CEN/TC 210	–	–	Under development
GRP tanks and vessels for use above ground – Part 1: Raw materials – Acceptance conditions and usage conditions	210006	prEN 13121-1	97/23/EC	CEN/TC 210	–	00:00:00	Under approval
GRP tanks and vessels for use above ground – Part 2: Composite materials – Chemical resistance	210007	prEN 13121-2	97/23/EC	CEN/TC 210	–	00:00:00	Under approval
GRP tanks and vessels for use above ground – Part 3: Design and workmanship	210008	–	97/23/EC	CEN/TC 210	–	–	Under development
GRP tanks and vessels for use above ground – Part 4: Delivery, installation and maintenance	210009	prEN 13121-4	97/23/EC	CEN/TC 210	–	00:00:00	Under approval

CEN/TC 267 (Industrial piping and pipelines)

Subject	Work Item ID	Standard Reference	Directive(s)	Technical Body	Start approval	Ratified	Current Status
Industrial piping and pipelines – Fibre reinforced plastic and plastic piping – Buried piping – Part 3: Safety devices	267023	–	97/23/EC	CEN/TC 267	–	–	Under development

Subject	Work item ID	Standard reference	Directive(s)	Technical Body	Start approval	Ratified	Current status
Metallic industrial piping – Part 1: General	267001	prEN 13480-1	97/23/EC	CEN/TC 267	–	00:00:00	Under approval
Metallic industrial piping – Part 2: Materials	267007	prEN 13480-2	97/23/EC	CEN/TC 267	–	00:00:00	Under approval
Metallic industrial piping – Part 3: Design and calculation	267008	prEN 13480-3	97/23/EC	CEN/TC 267	–	00:00:00	Under approval
Metallic industrial piping – Part 4: Fabrication and installation	267009	prEN 13480-4	97/23/EC	CEN/TC 267	–	00:00:00	Under approval
Metallic industrial piping – Part 5: Inspection and testing	267010	prEN 13480-5	97/23/EC	CEN/TC 267	–	00:00:00	Under approval
Metallic industrial piping – Part 6: Safety systems	267011	prEN 13480-6	97/23/EC	CEN/TC 267	–	00:00:00	Under approval
Metallic industrial piping – Part 7: Additional requirements for buried piping	267027	–	97/23/EC	CEN/TC 267	–	–	Under development

CEN/TC 69 (Industrial valves)

Subject	Work item ID	Standard reference	Directive(s)	Technical Body	Start approval	Ratified	Current status
Automatic steam traps – Classification (ISO 6704: 1982)	69005	EN 26704: 1991	97/23/EC	CEN/TC 69	–	30/09/1991	Ratified
Industrial valves – Steel check valves	69006	EN 26948: 1991	97/23/EC	CEN/TC 69	–	30/09/1991	Ratified
Industrial process control valves	69028	EN 1349: 2000	97/23/EC	CEN/TC 69	–	08/07/1999	Ratified
Industrial valves – Ball valves of thermoplastic materials	69070	prEN ISO 16135	97/23/EC	CEN/TC 69	–	00:00:00	Under approval

Industrial valves – Butt welding ends for steel valves	69080	EN 12627: 1999	97/23/EC	CEN/TC 69	–	16/04/1999	Ratified
Industrial valves – Butterfly valves of thermoplastic materials	69072	prEN ISO 16136	97/23/EC	CEN/TC 69	–	00:00:00	Under approval
Industrial valves – Cast iron check valves	69042	prEN 12334	97/23/EC	CEN/TC 69	–	00:00:00	Under approval
Industrial valves – Cast iron gate valves	69040	prEN 1171	97/23/EC	CEN/TC 69	–	00:00:00	Under approval
Industrial valves – Cast iron globe valves	69045	prEN 13789	97/23/EC	CEN/TC 69	–	00:00:00	Under approval
Industrial valves – Cast iron plug valves	69022	prEN 12335	97/23/EC	CEN/TC 69	–	00:00:00	Under approval
Industrial valves – Check valves made of thermoplastic materials	69074	prEN ISO 16137	97/23/EC	CEN/TC 69	–	00:00:00	Under approval
Industrial valves – Copper alloy ball valves	69062	prEN 13547	97/23/EC	CEN/TC 69	–	00:00:00	Under approval
Industrial valves – Copper alloy ball valves	69051	prEN 12328	97/23/EC	CEN/TC 69	–	00:00:00	Under approval
Industrial valves – Copper alloy gate valves (ANSI/ASME B1-20-1)	69050	prEN 12288	97/23/EC	CEN/TC 69	–	00:00:00	Under approval
Industrial valves – Copper alloy globe valves (ANSI/ASME B1-20-1)	69046	prEN 12360	97/23/EC	CEN/TC 69	–	00:00:00	Under approval
Industrial valves – Diaphragm valves made of metallic materials	69023	prEN 13397	97/23/EC	CEN/TC 69	–	00:00:00	Under approval
Industrial valves – Diaphragm valves of thermoplastic materials	69071	prEN ISO 16138	97/23/EC	CEN/TC 69	–	00:00:00	Under approval

Standards and Directives – Current Status

Description	Number	Reference	Directive	Committee		Date	Status
Industrial valves – End-to-end and centre-to-end dimensions for butt welding end valves	69010	EN 12982: 2000	97/23/EC	CEN/TC 69	–	06/11/1999	Ratified
Industrial valves – End-to-end and centre-to-end dimensions for butt welding end valves	69052	EN 558-2: 1995	97/23/EC	CEN/TC 69	–	16/10/1995	Ratified
Industrial valves – Gate valves of thermoplastic materials (ISO/DIS 16139: 2000)	69089	prEN ISO 16139	97/23/EC	CEN/TC 69	–	00:00:00	Under approval
Industrial valves – Globe valves made of thermoplastic materials	69073	–	97/23/EC	CEN/TC 69	–	–	Under development
Industrial valves – Isolating valves for LNG – Specification for suitability and appropriate verification tests	69049	prEN 12567	97/23/EC	CEN/TC 69	–	00:00:00	Under approval
Industrial valves – Isolating valves for LNG – Specification for suitability and appropriate verification tests	69020	EN 593: 1998	97/23/EC	CEN/TC 69	–	11/12/1997	Ratified
Industrial valves – Method for sizing the operating element	69077	EN 12570: 2000	97/23/EC	CEN/TC 69	–	09/04/2000	Ratified
Industrial valves – Method for sizing the operating element	69001	EN ISO 5210: 1996	97/23/EC	CEN/TC 69	–	20/10/1995	Ratified
Industrial valves – Part-turn valve actuator attachments (ISO/FDIS 5211: 2000)	69002	prEN ISO 5211	97/23/EC	CEN/TC 69	–	00:00:00	Under approval
Industrial valves – Part-turn valve actuator attachments (ISO/FDIS 5211: 2000)	69082	–	97/23/EC	CEN/TC 69	–	–	Under development

Subject	Work item ID	Standard reference	Directive(s)	Technical body	Start approval	Ratified	Current status
Industrial valves – Shell design strength – Part 2: Calculation methods for steel valves	69083	–	97/23/EC	CEN/TC 69	–	–	Under development
Industrial valves – Steel ball valves	69061	prEN 1983	97/23/EC	CEN/TC 69	–	00:00:00	Under approval
Industrial valves – Steel check valves	69043	–	97/23/EC	CEN/TC 69	–	–	Under development
Industrial valves – Steel check valves	69041	EN 1984: 2000	97/23/EC	CEN/TC 69	–	06/11/1999	Ratified

CEN/TC 269 (Shell and water tube boilers)

Subject	Work item ID	Standard reference	Directive(s)	Technical body	Start approval	Ratified	Current status
Shell boilers – Part 10: Requirements for feedwater and boiler water quality	269011	prEN 12953-10	97/23/EC	CEN/TC 269	–	00:00:00	Under approval
Shell boilers – Part 11: Acceptance tests	269027	–	97/23/EC	CEN/TC 269	–	–	Under development
Shell boilers – Part 12: Requirements for grate firing systems for solid fuels for the boiler	269028	–	97/23/EC	CEN/TC 269	–	–	Under development
Shell boilers – Part 13: Special requirements for stainless steel boiler servicing sterilizer	269031	–	97/23/EC	CEN/TC 269	–	–	Under development
Shell boilers – Part 1: General	269003	prEN 12953-1	97/23/EC	CEN/TC 269	–	00:00:00	Under approval

Standards and Directives – Current Status 361

Description	Number	Standard	Directive	TC		Date	Status
Shell boilers – Part 2: Materials for pressure parts of boilers and accessories	269004	prEN 12953-2	97/23/EC	CEN/TC 269	–	00:00:00	Under approval
Shell boilers – Part 3: Design and calculation for pressure parts	269005	prEN 12953-3	97/23/EC	CEN/TC 269	–	00:00:00	Under approval
Shell boilers – Part 4: Workmanship and construction of pressure parts of the boiler	269006	prEN 12953-4	97/23/EC	CEN/TC 269	–	00:00:00	Under approval
Shell boilers – Part 5: Inspection during construction, documentation and marking of pressure parts of the boiler	269012	prEN 12953-5	97/23/EC	CEN/TC 269	–	00:00:00	Under approval
Shell boilers – Part 6: Requirements for equipment for the boiler	269007	prEN 12953-6	97/23/EC	CEN/TC 269	–	00:00:00	Under approval
Shell boilers – Part 7: Requirements for firing systems for liquid and gaseous fuels	269008	prEN 12953-7	97/23/EC	CEN/TC 269	–	00:00:00	Under approval
Shell boilers – Part 8: Requirements for safeguards against excessive pressure	269009	prEN 12953-8	97/23/EC	CEN/TC 269	–	00:00:00	Under approval
Shell boilers – Part 9: Requirements for limiting devices and safety circuits of the boiler and accessories	269010	prEN 12953-9	97/23/EC	CEN/TC 269	–	00:00:00	Under approval
Water-tube boilers – Part 12: Requirements for feedwater and boiler water quality	269022	prEN 12952-12	97/23/EC	CEN/TC 269	–	00:00:00	Under approval
Water-tube boilers – Part 13: Requirements for flue gas cleaning systems	269023	prEN 12952-13	97/23/EC	CEN/TC 269	–	00:00:00	Under approval

ECISS/TC 31 (Steel castings)

Subject	Work item ID	Standard reference	Directive(s)	Technical body	Start approval	Ratified	Current status
Corrosion resistant steel castings	EC031008	EN 10283: 1998	97/23/EC	ECISS/TC 31	–	04/09/1998	Ratified
Founding – Technical conditions of delivery – Part 2: Additional requirements for steel castings	EC031005	EN 1559-2: 2000	97/23/EC	ECISS/TC 31	–	03/01/2000	Ratified
Heat resistant steel castings	EC031009	prEN 10295	97/23/EC	ECISS/TC 31	–	00:00:00	Under approval
Steel castings – Welding homologation procedure	EC031006	–	97/23/EC	ECISS/TC 31	–	–	Under development
Steel castings for structural and general engineering uses – Part 2: Steel castings for structural uses	EC031010	prEN 10293-2	97/23/EC	ECISS/TC 31	–	00:00:00	Under approval
Steel castings for structural and general engineering uses – Part 3: Steel castings for general engineering uses	EC031011	prEN 10293-3	97/23/EC	ECISS/TC 31	–	00:00:00	Under approval
Technical delivery conditions for steel castings for pressure purposes – Part 1: General	EC031001	EN 10213-1: 1995	97/23/EC	ECISS/TC 31	–	20/10/1995	Ratified
Technical delivery conditions for steel castings for pressure purposes – Part 2: Steel grades for use at room temperature and elevated temperatures	EC031002	EN 10213-2: 1995	97/23/EC	ECISS/TC 31	–	20/10/1995	Ratified

Subject	Work item ID	Standard reference	Directive(s)	Technical body	Start approval	Ratified	Current status
Technical delivery conditions for steel castings for pressure purposes – Part 3: Steel grades for use at low temperatures	EC031003	EN 10213-3: 1995	97/23/EC	ECISS/TC 31	–	20/10/1995	Ratified
Technical delivery conditions for steel castings for pressure purposes – Part 4: Austenitic and austenitic-ferritic steel grades	EC031004	EN 10213-4: 1995	97/23/EC	ECISS/TC 31	–	20/10/1995	Ratified

ECISS/TC 28 (Steel forgings)

Subject	Work item ID	Standard reference	Directive(s)	Technical body	Start approval	Ratified	Current status
Non-destructive testing of steel forgings – Part 1: Magnetic particle inspection	EC028020	EN 10228-1: 1999	97/23/EC	ECISS/TC 28	–	01/03/1999	Ratified
Non-destructive testing of steel forgings – Part 2: Penetrant testing	EC028021	EN 10228-2: 1998	97/23/EC	ECISS/TC 28	–	21/12/1997	Ratified
Non-destructive testing of steel forgings – Part 3: Ultrasonic testing of ferritic or martensitic steel forgings	EC028027	EN 10228-3: 1998	97/23/EC	ECISS/TC 28	–	21/12/1997	Ratified
Non-destructive testing of steel forgings – Part 4: Ultrasonic testing of austenitic and austenitic-ferritic stainless steel forgings	EC028026	EN 10228-4: 1999	97/23/EC	ECISS/TC 28	–	09/07/1999	Ratified
Open die steel forgings for general engineering purposes – Part 1: General requirements	EC028012	EN 10250-1: 1999	97/23/EC	ECISS/TC 28	–	16/07/1999	Ratified

Description	Code	Standard	Directive	Committee		Date	Status
Open die steel forgings for general engineering purposes – Part 2: Non-alloy quality and special steels	EC028013	EN 10250-2: 1999	97/23/EC	ECISS/TC 28	–	09/09/1999	Ratified
Open die steel forgings for general engineering purposes – Part 3: Alloy special steels	EC028014	EN 10250-3: 1999	97/23/EC	ECISS/TC 28	–	09/09/1999	Ratified
Open die steel forgings for general engineering purposes – Part 4: Stainless steels	EC028016	EN 10250-4: 1999	97/23/EC	ECISS/TC 28	–	09/09/1999	Ratified
Semi-finished products for forging – Tolerances on dimensions, shape and weight	EC028029	prEN 10031	97/23/EC	ECISS/TC 28	–	00:00:00	Under approval
Steel closed die forgings – General technical delivery conditions	EC028025	EN 10254: 1999	97/23/EC	ECISS/TC 28	–	22/08/1999	Ratified
Steel die forgings – Tolerances on dimensions – Part 1: Drop and vertical press forgings	EC028023	EN 10243-1: 1999	97/23/EC	ECISS/TC 28	–	22/08/1999	Ratified
Steel die forgings – Tolerances on dimensions – Part 2: Upset forging made on horizontal forging machines	EC028004	EN 10243-2: 1999	97/23/EC	ECISS/TC 28	–	22/08/1999	Ratified
Steel forgings for pressure purposes – Part 1: General requirements for open die forgings	EC028007	EN 10222-1: 1998	97/23/EC	ECISS/TC 28	–	26/10/1997	Ratified
Steel forgings for pressure purposes – Part 2: Ferritic and martensitic steels with specified elevated temperature properties	EC028008	EN 10222-2: 1999	97/23/EC	ECISS/TC 28	–	05/09/1999	Ratified

ECISS/TC 29 (Steel tubes and fittings for steel tubes)

Subject	Work item ID	Standard reference	Directive(s)	Technical body	Start approval	Ratified	Current status
Butt welding pipe fittings – Part 2: Wrought carbon and ferritic alloy steels with specific inspection requirements	EC029026	prEN 10253-2	97/23/EC	ECISS/TC 29	–	00:00:00	Under approval
Corrugated flexible metallic hose and hose assemblies (ISO 10380: 1994 rev.)	EC029078	–	97/23/EC	ECISS/TC 29	–	–	Under development
Internal and/or external protective coatings for steel tubes – Specification for hot dip galvanized coatings applied in automatic plants	EC029015	EN 10240: 1997	97/23/EC	ECISS/TC 29	–	09/08/1997	Ratified
Malleable cast iron fittings with compression ends for polyethylene (PE) piping systems	EC029057	EN 10284: 2000	97/23/EC	ECISS/TC 29	–	12/11/1999	Ratified
Metallic materials – Tube (in full section) – Bend test	EC029006	EN 10232: 1993	97/23/EC	ECISS/TC 29	–	25/10/1993	Ratified
Metallic materials – Tube – Drift expanding test	EC029008	EN 10234: 1993	97/23/EC	ECISS/TC 29	–	25/10/1993	Ratified
Metallic materials – Tube – Flanging test	EC029009	EN 10235: 1993	97/23/EC	ECISS/TC 29	–	25/10/1993	Ratified
Metallic materials – Tube – Flattening test	EC029007	EN 10233: 1993	97/23/EC	ECISS/TC 29	–	25/10/1993	Ratified
Metallic materials – Tube – Ring expanding test	EC029010	EN 10236: 1993	97/23/EC	ECISS/TC 29	–	25/10/1993	Ratified

Description	Code	Standard	Directive	Committee		Date	Status
Metallic materials – Tube – Ring tensile test	EC029019	EN 10237: 1993	97/23/EC	ECISS/TC 29	–	25/10/1993	Ratified
Metallic materials – Tube – Ring tensile test	EC029030	EN 10275: 1999	97/23/EC	ECISS/TC 29	–	16/04/1999	Ratified
Metallic tube connections for fluid power and general use – Part 1: 24° compression fittings (ISO 8434-1: 1994)	EC029058	EN ISO 8434-1: 1997	97/23/EC	ECISS/TC 29	–	16/02/1996	Ratified
Non destructive testing of steel tubes – Part 1: Automatic electromagnetic testing of seamless and welded (except submerged arc welded) ferromagnetic steel tubes for verification of hydraulic leak-tightness	EC029011	EN 10246-1: 1996	97/23/EC	ECISS/TC 29	–	28/12/1995	Ratified
Non destructive testing of steel tubes – Part 7: Automatic full peripheral ultrasonic testing of seamless and welded (except submerged arc welded) steel tubes for the detection of longitudinal imperfections	EC029012	EN 10246-7: 1996	97/23/EC	ECISS/TC 29	–	28/12/1995	Ratified
Non-destructive testing of steel tubes – Part 10: Radiographic testing of the weld seam of automatic fusion arc welded steel tubes for the detection of imperfections	EC029075	EN 10246-10: 2000	97/23/EC	ECISS/TC 29	–	29/03/2000	Ratified

Description	Code	Standard	Directive	Committee		Date	Status
Non-destructive testing of steel tubes – Part 11: Liquid penetrant testing of seamless and welded steel tubes for the detection of surface imperfections	EC029074	EN 10246-11: 2000	97/23/EC	ECISS/TC 29	–	25/12/1999	Ratified
Non-destructive testing of steel tubes – Part 12: Magnetic particle inspection of seamless and welded ferromagnetic steel tubes for the detection of surface imperfections	EC029069	EN 10246-12: 2000	97/23/EC	ECISS/TC 29	–	25/12/1999	Ratified
Non-destructive testing of steel tubes – Part 13: Automatic full peripheral ultrasonic thickness testing for seamless and welded (except submerged arc welded) steel tubes	EC029071	EN 10246-13: 2000	97/23/EC	ECISS/TC 29	–	25/12/1999	Ratified
Non-destructive testing of steel tubes – Part 14: Automatic ultrasonic testing of seamless and welded (except submerged arc-welded) steel tubes for the detection of laminar imperfections	EC029036	EN 10246-14: 1999	97/23/EC	ECISS/TC 29	–	06/10/1999	Ratified
Non-destructive testing of steel tubes – Part 15: Automatic ultrasonic testing of strip/plate used in the manufacture of welded steel tubes for the detection of laminar imperfections	EC029073	EN 10246-15: 2000	97/23/EC	ECISS/TC 29	–	25/12/1999	Ratified

Description	Ref	Standard	Directive	Committee		Date	Status
Non-destructive testing of steel tubes – Part 16: Automatic ultrasonic testing of the area adjacent to the weld seam of welded steel tubes for the detection of laminar imperfections	EC029070	EN 10246-16: 2000	97/23/EC	ECISS/TC 29	—	25/12/1999	Ratified
Non-destructive testing of steel tubes – Part 17: Ultrasonic testing of tube ends of seamless and welded steel tubes for the detection of laminar imperfections	EC029072	EN 10246-17: 2000	97/23/EC	ECISS/TC 29	—	25/12/1999	Ratified
Non-destructive testing of steel tubes – Part 18: Magnetic particle inspection of tube ends of seamless and welded ferromagnetic steel tubes for the detection of laminar imperfections	EC029068	EN 10246-18: 2000	97/23/EC	ECISS/TC 29	—	25/12/1999	Ratified
Non-destructive testing of steel tubes – Part 2: Automatic eddy current testing of seamless and welded (except submerged arc-welded) austenitic and austenitic-ferritic steel tubes for verification of hydraulic leak-tightness	EC029052	EN 10246-2: 2000	97/23/EC	ECISS/TC 29	—	25/12/1999	Ratified
Non-destructive testing of steel tubes – Part 3: Automatic eddy current testing of seamless and welded (except submerged arc-welded) steel tubes for the detection of imperfections	EC029037	EN 10246-3: 1999	97/23/EC	ECISS/TC 29	—	06/10/1999	Ratified

Non-destructive testing of steel tubes – Part 4: Automatic full peripheral magnetic transducer/flux leakage testing of seamless ferromagnetic steel tubes for the detection of transverse imperfections	EC029040	EN 10246-4: 1999	97/23/EC	ECISS/TC 29	–	06/10/1999	Ratified
Non-destructive testing of steel tubes – Part 5: Automatic full peripheral magnetic transducer/flux leakage testing of seamless and welded (except submerged arc welded) ferromagnetic steel tubes for the detection of longitudinal imperfections	EC029039	EN 10246-5: 1999	97/23/EC	ECISS/TC 29	–	06/10/1999	Ratified
Non-destructive testing of steel tubes – Part 6: Automatic full peripheral ultrasonic testing of seamless steel tubes for the detection of transverse imperfections	EC029038	EN 10246-6: 1999	97/23/EC	ECISS/TC 29	–	06/10/1999	Ratified
Non-destructive testing of steel tubes – Part 8: Automatic ultrasonic testing of the weld seam of electric welded steel tubes for the detection of longitudinal imperfections	EC029041	EN 10246-8: 1999	97/23/EC	ECISS/TC 29	–	06/10/1999	Ratified

Description	Code	Standard	Directive	Committee	Date	Status
Non-destructive testing of steel tubes – Part 9: Automatic ultrasonic testing of the weld seam of submerged arc welded steel tubes for the detection of longitudinal and/or transverse imperfections	EC029035	EN 10246-9: 2000	97/23/EC	ECISS/TC 29	–	25/12/1999 Ratified
Non-destructive testing of steel tubes – Qualification and competence of level 1 and 2 non-destructive testing personnel	EC029067	EN 10256: 2000	97/23/EC	ECISS/TC 29	–	29/03/2000 Ratified
Pipe threads where pressure tight joints are made on the threads – Part 1: Designation, dimensions and tolerances	EC029004	prEN 10226-1	97/23/EC	ECISS/TC 29	–	00:00:00 Under approval
Pipe threads where pressure tight joints are made on the threads – Part 2: Internal conical thread	EC029096	–	97/23/EC	ECISS/TC 29	–	– Under development
Pipe threads where pressure tight joints are made on the threads – Verification by means of limit gauges	EC029097	prEN ISO 7-2	97/23/EC	ECISS/TC 29	–	00:00:00 Under approval
Pipe threads where pressure-tight joints are not made on the threads – Part 1: Designation, dimensions and tolerances (Revision of ISO 228-1: 1994)	EC029005	prEN ISO 228-1	97/23/EC	ECISS/TC 29	–	00:00:00 Under approval

Standards and Directives – Current Status 373

Pipework – Corrugated flexible metallic hose assemblies for the protection of electrical cables in explosive atmosphere (ISO 10807: 1994)	EC029080	EN ISO 10807: 1996	97/23/EC	ECISS/TC 29	–	06/01/1996	Ratified
Pipework – Flexible metallic hoses – Vocabulary of general terms	EC029077	ISO/DIS 7369	97/23/EC	ECISS/TC 29	–	00:00:00	Under approval
Pipework – Non-alloyed and stainless steel fittings for corrugated flexible metallic hoses (ISO 10806: 1994)	EC029079	–	97/23/EC	ECISS/TC 29	–	–	Under development
Pipework – Non-alloyed and stainless steel fittings for corrugated flexible metallic hoses (ISO 10806: 1994)	EC029018	ENV 10220: 1993	97/23/EC	ECISS/TC 29	–	05/11/1993	Ratified
Seamless steel tubes for pressure purposes – Technical delivery conditions – Part 1: Non-alloy steel tubes with specified room temperature properties	EC029013	prEN 10216-1	97/23/EC	ECISS/TC 29	–	00:00:00	Under approval
Seamless steel tubes for pressure purposes – Technical delivery conditions – Part 2: Non-alloy and alloy steel tubes with specified elevated temperature properties	EC029024	prEN 10216-2	97/23/EC	ECISS/TC 29	–	00:00:00	Under approval
Seamless steel tubes for pressure purposes – Technical delivery conditions – Part 2: Non-alloy and alloy steel tubes with specified elevated temperature properties	EC029055	prEN 10216-3	97/23/EC	ECISS/TC 29	–	00:00:00	Under approval

Description	Code	Standard	Directive	Committee		Date	Status
Seamless steel tubes for pressure purposes – Technical delivery conditions – Part 4: Non-alloy and alloy steel tubes with specified low temperature properties	EC029046	prEN 10216-4	97/23/EC	ECISS/TC 29	–	00:00:00	Under approval
Seamless steel tubes for pressure purposes – Technical delivery conditions – Part 5: Stainless steel tubes	EC029053	prEN 10216-5	97/23/EC	ECISS/TC 29	–	00:00:00	Under approval
Stainless steel tubes – Dimensions, tolerances and conventional masses per unit length (ISO 1127: 1992)	EC029066	EN ISO 1127: 1996	97/23/EC	ECISS/TC 29	–	16/02/1996	Ratified
Steel threaded pipe fittings	EC029027	EN 10241: 2000	97/23/EC	ECISS/TC 29	–	29/03/2000	Ratified
Steel tubes for precision applications – Technical delivery conditions – Part 1: Seamless cold drawn tubes	EC029088	prEN 10305-1	97/23/EC	ECISS/TC 29	–	00:00:00	Under approval
Steel tubes for precision applications – Technical delivery conditions – Part 4: Seamless cold drawn tubes for hydraulic and pneumatic power systems	EC029099	prEN 10305-4	97/23/EC	ECISS/TC 29	–	00:00:00	Under approval
Steel tubes, fittings and structural hollow sections – Definitions and symbols for use in product standards	EC029043	prEN 10266	97/23/EC	ECISS/TC 29	–	00:00:00	Under approval

Steel tubes, fittings and structural hollow sections – Definitions and symbols for use in product standards	EC029022	EN 10242: 1994	97/23/EC	ECISS/TC 29	–	10/11/1994	Ratified
Welded steel tubes for pressure purposes – Technical delivery conditions – Part 1: Non-alloy steel tubes with specified room temperature properties	EC029014	prEN 10217-1	97/23/EC	ECISS/TC 29	–	00:00:00	Under approval
Welded steel tubes for pressure purposes – Technical delivery conditions – Part 2: Electric welded non-alloy and alloy steel tubes with specified elevated temperature properties	EC029034	prEN 10217-2	97/23/EC	ECISS/TC 29	–	00:00:00	Under approval
Welded steel tubes for pressure purposes – Technical delivery conditions – Part 3: Alloy fine grain steel tubes	EC029056	prEN 10217-3	97/23/EC	ECISS/TC 29	–	00:00:00	Under approval
Welded steel tubes for pressure purposes – Technical delivery conditions – Part 4: Electric welded non-alloy steel tubes with specified low temperature properties	EC029047	prEN 10217-4	97/23/EC	ECISS/TC 29	–	00:00:00	Under approval
Welded steel tubes for pressure purposes – Technical delivery conditions – Part 5: Submerged arc welded non-alloy and alloy steel tubes with specified elevated temperature properties	EC029048	prEN 10217-5	97/23/EC	ECISS/TC 29	–	00:00:00	Under approval

Subject						
Welded steel tubes for pressure purposes – Technical delivery conditions – Part 6: Submerged arc welded non-alloy steel tubes with specified low temperature properties	EC029049	prEN 10217-6	97/23/EC	ECISS/TC 29	—	00:00:00	Under approval
Welded steel tubes for pressure purposes – Technical delivery conditions – Part 7: Stainless steel tubes	EC029054	prEN 10217-7	97/23/EC	ECISS/TC 29	—	00:00:00	Under approval

ECISS/TC 22 (Steels for pressure purposes - Qualities)

Subject	Work item ID	Standard reference	Directive(s)	Technical body	Start approval	Ratified	Current status
Flat products made of steels for pressure purposes – Part 1: General requirements	EC022017	EN 10028-1: 2000	97/23/EC	ECISS/TC 22	—	29/10/1999	Ratified
Flat products made of steels for pressure purposes – Part 2: Non-alloy and alloy steels with specified elevated temperature properties	EC022002	EN 10028-2: 1992	97/23/EC	ECISS/TC 22	—	21/12/1992	Ratified
Flat products made of steels for pressure purposes – Part 3: Weldable fine grain steels, normalized	EC022003	EN 10028-3: 1992	97/23/EC	ECISS/TC 22	—	21/12/1992	Ratified
Flat products made of steels for pressure purposes – Part 4: Nickel alloy steels with specified low temperature properties	EC022011	EN 10028-4: 1994	97/23/EC	ECISS/TC 22	—	16/09/1994	Ratified

Description	Code	Standard	Directive	Committee		Date	Status
Flat products made of steels for pressure purposes – Part 5: Weldable fine grain steels, thermomechanically rolled	EC022012	EN 10028-5: 1996	97/23/EC	ECISS/TC 22	–	14/11/1996	Ratified
Flat products made of steels for pressure purposes – Part 6: Weldable fine grain steels, quenched and tempered	EC022013	EN 10028-6: 1996	97/23/EC	ECISS/TC 22	–	19/11/1996	Ratified
Flat products made of steels for pressure purposes – Part 7: Stainless steels	EC022016	EN 10028-7: 2000	97/23/EC	ECISS/TC 22	–	03/09/1999	Ratified
Hot rolled weldable steel bars for pressure purposes with specified elevated temperature properties	EC022019	EN 10273: 2000	97/23/EC	ECISS/TC 22	–	29/10/1999	Ratified
Method for the derivation of minimum values of proof strength of steel at elevated temperatures	EC022014	prEN 10314	97/23/EC	ECISS/TC 22	–	00:00:00	Under approval
Stainless steel bars for pressure purposes	EC022020	prEN 10272	97/23/EC	ECISS/TC 22	–	00:00:00	Under approval
Steel products for pressure purposes – Derivation and verification of elevated temperature properties – Part 1: Yield or proof stress of carbon and low alloy steel products (ISO 2605-1:1976)	EC022006	ENV 22605-1: 1991	97/23/EC	ECISS/TC 22	–	16/10/1991	Ratified
Steel products for pressure purposes – Derivation and verification of elevated temperature properties – Part 2: Proof stress of austenitic steel products (ISO 2605-2:1976)	EC022007	ENV 22605-2: 1991	97/23/EC	ECISS/TC 22	–	16/10/1991	Ratified

Steel products for pressure purposes – Derivation and verification of elevated temperature properties – Part 3: An alternative procedure for deriving the elevated temperature yield or proof stress properties when data are limited (ISO 2605-3:1985)	EC022008	ENV 22605-3: 1991	97/23/EC	ECISS/TC 22	–	16/10/1991 Ratified
Steel sheet and strip for welded gas cylinders	EC022015	EN 10120: 1996	97/23/EC	ECISS/TC 22	–	14/11/1996 Ratified
Steels and nickel alloys for fasteners with specified elevated and/or low temperature properties	EC022010	EN 10269: 1999	97/23/EC	ECISS/TC 22	–	01/07/1999 Ratified
Steels for simple pressure vessels – Technical delivery requirements for plates, strips and bars	EC022005	EN 10207: 1991	97/23/EC	ECISS/TC 22	–	30/11/1991 Ratified

ECISS/TC 10 (Structural steels - Grades and qualities)

Subject	Work item ID	Standard reference	Directive(s)	Technical body	Start approval	Ratified	Current status
Automatically blast-cleaned and automatically prefabricated primed structural steel products	EC010011	EN 10238: 1996	97/23/EC	ECISS/TC 10	–	19/03/1996	Ratified
Cold formed welded structural hollow sections of non-alloy and fine grain steels – Part 1: Technical delivery requirements	EC010019	EN 10219-1: 1997	97/23/EC	ECISS/TC 10	–	22/06/1997	Ratified

Standards and Directives – Current Status

Cold formed welded structural hollow sections of non-alloy and fine grain steels – Part 2: Tolerances, dimensions and sectional properties	EC010020	EN 10219-2: 1997	97/23/EC	ECISS/TC 10	–	22/06/1997 Ratified
Delivery requirements for surface condition of hot rolled steel plates and sections – Part 3: Sections	EC010005	EN 10163-3: 1991	97/23/EC	ECISS/TC 10	–	21/08/1991 Ratified
Delivery requirements for surface condition of hot rolled steel plates, wide flats and sections – Part 1: General requirements	EC010003	EN 10163-1: 1991	97/23/EC	ECISS/TC 10	–	21/08/1991 Ratified
Delivery requirements for surface condition of hot rolled steel plates, wide flats and sections – Part 2: Plate and wide flats	EC010012	EN 10163-2: 1991	97/23/EC	ECISS/TC 10	–	21/08/1991 Ratified
Hot finished structural hollow sections of non-alloy and fine grain structural steels – Part 1: Technical delivery requirements	EC010009	EN 10210-1: 1994	97/23/EC	ECISS/TC 10	–	07/03/1994 Ratified
Hot finished structural hollow sections of non-alloy and fine grain structural steels – Part 2: Tolerances, dimensions and sectional properties	EC010006	EN 10210-2: 1997	97/23/EC	ECISS/TC 10	–	22/06/1997 Ratified
Hot finished structural hollow sections of non-alloy and fine grain structural steels – Part 2: Tolerances, dimensions and sectional properties	EC010033	prEN 10025-1	97/23/EC	ECISS/TC 10	–	00:00:00 Under approval

Description	Code	Standard	Directive	Committee		Date	Status
Hot rolled products of non-alloy structural steels – Technical delivery conditions	EC010001	EN 10025: 1990	97/23/EC 89/106/EEC	ECISS/TC 10	—	30/03/1990	Ratified
Hot-rolled flat products made of high yield strength steels for cold forming – Part 1: General delivery conditions	EC010024	EN 10149-1: 1995	97/23/EC	ECISS/TC 10	—	06/08/1995	Ratified
Hot-rolled flat products made of high yield strength steels for cold forming – Part 2: Delivery conditions for thermomechanically rolled steels	EC010028	EN 10149-2: 1995	97/23/EC	ECISS/TC 10	—	06/08/1995	Ratified
Hot-rolled flat products made of high yield strength steels for cold forming – Part 3: Delivery conditions for normalized or normalized rolled steels	EC010027	EN 10149-3: 1995	97/23/EC	ECISS/TC 10	—	06/08/1995	Ratified
Hot-rolled products in weldable fine grain structural steels – Part 1: General delivery conditions	EC010002	EN 10113-1: 1993	97/23/EC	ECISS/TC 10	—	05/03/1993	Ratified
Hot-rolled products in weldable fine grain structural steels – Part 2: Delivery conditions for normalized/normalized rolled steels	EC010018	EN 10113-2: 1993	97/23/EC	ECISS/TC 10	—	05/03/1993	Ratified

GAS CYLINDERS

Subject	Work item ID	Standard reference	Directive(s)	Technical body	Start approval	Ratified	Current status
Pressure regulators integrated with cylinder valves	23091	–	97/23/EC	CEN/TC 23	–	–	Under development
Seamless aluminium alloy cylinders for carbon dioxide fire extinguishers – Design, manufacture and testing	23088	–	97/23/EC	CEN/TC 23	–	–	Under development
Transportable gas cylinders – Seamless steel gas cylinders for portable extinguishers and bottles for breathing apparatus with a water capacity of not more than 15 litres	23092	–	97/23/EC	CEN/TC 23	–	–	Under development

CEN/TC 54 (Unfired pressure vessels)

Subject	Work item ID	Standard reference	Directive(s)	Technical body	Start approval	Ratified	Current status
Closed expansion vessels with built-in diaphragm for installation in water systems	54010	prEN 13831	97/23/EC	CEN/TC 54	–	00:00:00	Under approval
Gas-loaded accumulators for fluid power applications	54019	-	97/23/EC	CEN/TC 54	–	–	Under development
Gas-loaded accumulators for fluid power applications	54023	-	97/23/EC	CEN/TC 54	–	–	Under development

Gas-loaded accumulators for fluid power applications	54020	prEN 764-2	97/23/EC	CEN/TC 54	–	00:00:00	Under approval
Pressure equipment – Part 3: Definition and parties involved	54021	prEN 764-3	97/23/EC	CEN/TC 54	–	00:00:00	Under approval
Pressure equipment – Part 4: Establishment of technical delivery conditions for materials	54022	prEN 764-4	97/23/EC	CEN/TC 54	–	00:00:00	Under approval
Pressure equipment – Part 5: Compliance and inspection documentation of materials	54025	prEN 764-5	97/23/EC	CEN/TC 54	–	00:00:00	Under approval
Pressure equipment – Part 6: Operating instructions	54024	–	97/23/EC	CEN/TC 54	–	–	Under development
Unfired pressure vessels – Part 1: General	54007	prEN 13445-1	97/23/EC	CEN/TC 54	–	00:00:00	Under approval
Unfired pressure vessels – Part 2: Materials	54012	prEN 13445-2	97/23/EC	CEN/TC 54	–	00:00:00	Under approval
Unfired pressure vessels – Part 3: Design	54013	prEN 13445-3	97/23/EC	CEN/TC 54	–	00:00:00	Under approval
Unfired pressure vessels – Part 4: Manufacture	54014	prEN 13445-4	97/23/EC	CEN/TC 54	–	00:00:00	Under approval
Unfired pressure vessels – Part 5: Inspection and testing	54015	prEN 13445-5	97/23/EC	CEN/TC 54	–	00:00:00	Under approval

Index

ASME A stamp 131
Accreditation 132
Acronyms, NDT 223
Air receivers 5, 98
  BS 5169 99
Allowable stresses 81
Alloy:
  nickel 189
  steels 185
All-weld tensile test 207
American welding symbols 200
ANSI B16.34 144, 147
ANSI B16.5 classes 137
ANSI B31.1 code limits 159
ANSI/FCI 70-2 150
API 598 150
Area replacement method 90
ASME 131
  certification 126
  code 254
  intent 95, 96
  vessel codes 76
  VIII 77, 109, 225
  VIII-1:
    allowable stress values 82
    basic equations 83
    welded joint categories 83
ASME/ANSI B31.1 code 158
ASTM hole-type penetrameter 222
Austenitic stainless steels 186
Authorized Inspector (AI) 126
Axial compression 84

Basic equations ASME VIII-1 83
Bend test 207
Boilers 4
  failure modes 241
  fire tube
  principles 179–180
  shell 101, 104, 361
  water-tube 107, 361
Bolted flanges 89
BS 113/pr EN 12952 107
BS 2790 101
BS EN 287 203
BS EN 288 203
BS welding symbols 198
Butt joint 'joggled' 43

Canadian standards association
  B51-97 106
Castings:
  steel 365
  UT of 214
Categories of SPVs 257
Caustic embrittlement 242
Caustic gouging 242
CE mark 255
CEN/CENELEC 250
Certificate of conformity 132
Certification 132
  vessel 'statutory' 253
Chemical corrosion 238
Circularity 122
Circumference 120
CODAP 95 107

Code:
  ASME intent 59
  compliance 59
  ASME intent 59
  symbol stamps 129
Compensation 29
Competent person 324
Compression, axial 84
Conformity:
  assessment 132
  declaration of 311
Contact-type deaerator 54
Contact-type exchangers 4
Corrosion 238
  allowance 23
  chemical 238
  crevice 240
  electrolytic 238
  erosion 240
  fatigue 240, 243
  galvanic 238, 239
  intergranular 240
  stress 240
  waterwall 243
Covers 88
Creep 235, 237
  high-temperature 242
  rate 237
  recovery 237
  rupture strength 237
  strength 237
Crevice corrosion 240
Cylinders, external pressure on 85
Cylindrical shell theory 21
Cylindrical shells under internal
    pressure 84
Cylindrical vessel shells 81

DBA stress categories 56
Deaerator, contact-type 54
Declaration of conformity 311
Defect acceptance criteria, pressure
    vessel codes 226

Department of Trade and Industry 6
Design by Analysis (DBA) 53
Destructive testing of welds 205
Dimensional examinations 114
Directives 6
Discontinuities:
  non-planar 216
  planar 216
Dished ends:
  elliposoidal 87
  hemispherical 86
  torispherical 'shallow' 88
Distortion 118
Draft CEN/TC269 108
Drum swell 180
Duplex stainless steels 186
Dye penetrant (DP) testing 209

ASME E stamp 131
Electrolytic corrosion 238
Ellipsoid 27 *et seq*
Embrittlement, caustic 242
EN 12953 101, 105
EN 1442 41
EN 45004:1995 132
EN10 204 classes 191
Energy, internal 338
Enthalpy 167, 338
EPERC 325
Equivalent stress 27
Erosion corrosion 240
Essential Safety Requirements
    (ESRs) 251
  checklist 295
EU new approaches 250
European Directives 351 *et seq*
European harmonized standard 105
European Notified Bodies (PED)
    343 *et seq*
European Pressure Equipment
    Directive (PED) 131
European Pressure Equipment
    Research Council (EPERC) 325

Examinations:
  dimensional 114
  visual 114
Exfoliation 244
External pressure on cylinders 85

Failure 229
Fatigue 232
  analysis 95
  corrosion 243
  limits 234
  S–N curve 232
  strength, – rules of thumb 234
Feed heater, high-pressure 54
Ferritic stainless steels 186
Flanges 88, 137, 352
  bolted 89
  facings 140
  forger 5
  materials 138
  types 140
Flat plates 85, 88
Forced circulation 177
Forged flanges 5
Forgings, steel 367

Galvanic corrosion 238, 239
Gas characteristics 173
Gas cylinder 4, 265, 318, 384
  basic design 39 *et seq*
  ISO technical standards for 268
  marking for 46
Gouging, caustic 242
GRP tanks 356

Hardness text 208
Harmonized standards 6, 250
  SPV 264
Heat exchangers 4, 90
  basic design 47
Heat recovery steam generators
  (HRSGs) 173
Heat transfer 167

Hemispherical dished ends 86
High-pressure feed heater 54
High-temperature creep 242
Horizontal pressure vessels 34
HRSGs 4
  level control 179
  materials 183
  operation 176
Hydrogen attack 243
Hydrostatic test 123

Industrial valves 358
In-service inspection 321
Inspection 113, 132
  and test plans (ITPs) 60
  openings 32
Instrument identification 20
Intergranular corrosion 240
Internal energy 338
Internal pressure, thick cylindrical
    shells under 84
ISO 9000 133
ISO technical standards for gas
    cylinders 268

Joggled butt joint 43
Joint efficiency factors (E) 81

KKS:
  codes 9
  power plant classification 7

Latent heat 168
Leak:
  rate 125
  testing 122
LEFM method 230
Legislation 6

ASME M stamp 130
Macro test 208
Magnetic particle (MP)
    testing 212

Manufacturers' data report
   forms 127
Marking for gas cylinders 46
Martensitic stainless steels 186
Material:
   body-grades 214
   edge-grades 214
   standards 192
   traceability 190
Maximum principal stress
   theory 232
Metallographic replication 236
Misalignment 118
Multi-axis stress states 231

National Standards 251
Natural circulation 177
NDT acronyms 223
NDT vessel code applications 225
Nickel alloys 189
Non-destructive testing (NDT)
   techniques 209
Non-planar discontinuities 216
Notified body 132, 314
Notifying authority 132
Nozzle types 156

Pads 33
PD 5500 64, 70, 109, 225
   Cat 1 68
   Cat 2 69
Penetrameters 219
Pipeline designation system 164
Pipework classification 161
Plain carbon steels 185
Planar discontinuities 216
Pneumatic testing 124
Popping pressure 154
Power piping 158
Power plant classification, KKS 7
   et seq
PP stamp, ASME 131
pr EN 13445 53, 55

pr EN 12952 108
Pressure conversions 171
Pressure equipment
   categories 11
   directive (PED) 6, 248, 318
      97/23/EC 271
      97/23/EC 252
      conformity assessment
         procedures 275
      essential safety requirements
         70, 295
      risk categories 278
      structure 274
   legislation 249
   manufacturing methods 113
   marking 312
   regulations 318
      1999 312
      SI 1999/2001 252, 272
   symbols 13 *et seq*
   vessel nameplate 119
Pressure receptacles,
   transportable 4
Pressure Systems Safety
   Regulations 2000 321, 322
Pressure testing 122
Pressure vessel 385
   codes 59, 110
      defect acceptance criteria 226
   cylinders 24 *et seq*
   design 4
   ITP, typical content 61
   simple 4, 252, 255
   software 4
Primary stress category 93
Process and Instrumentation
   Diagrams (PIDs) 13 *et seq*
Profile 122
PSSRs 319

QA 113

Radiographic testing (RT) 219

Reinforcement 29
  limits 91
Relevant fluid 323
Replication, metallographic 236

ASME S stamp 129
Safety Assessment Federation
    (SAFeD) 133
Safety:
  accessories 274
  clearance 258
    for SPVs 259
  devices 151
  valves 178
    materials 152
    terminology 153
Scope of NDT 147
Shell boilers 101, 104, 361
  fittings, 102
Shrink effect 180
Side bend test 208
Simmer 155
Simple pressure vessels (SPVs) 4, 35, 252, 255
  calculations 37
  directive 256, 318
    87/404/EEC 252, 256
  harmonized standards 264
  regulations 256, 318
  safety clearance for 259
S–N curve, fatigue 232
Specific heat 167
Spherical shells 86
SPVs, categories of 257
Stainless steels 186
  austenitic 186
  duplex 186
  ferritic 186
  martensitic 186
Stamp:
  ASME A 131
  ASME E 131
  ASME M 130

ASME PP 131
ASME S 129
ASME V 131
Statutory instrument (SI) 6
  document 317
Steam:
  characteristics 168
  properties data 337 *et seq*
  temperature control 181
Steel:
  alloy 185
  austenitic stainless 186
  castings 365
  duplex stainless 186
  ferritic stainless 186
  forgings 367
  martensitic stainless 186
  plain carbon 185
  stainless 186
  structural 382
  tubes 369
Straightness 120
Stress:
  categories 92
  categorization 56
  combinations 95
  corrosion 240
  limits 56
  loading 233
  theory, maximum principal 232
  thermal 94
Structural discontinuity 93
Structural steels 382
Surface misalignment 120
Surface-type exchangers 47
  configurations 50

Temperature conversions 172
Tensile test 207
Testing 113
Thermal design 47
Thermal stress 94
  general 94

local 94
Thick cylindrical shells under internal pressure 84
Third-party organizations 343
Toleranced features 119
Torisphere 29
TPED 270
Transportable pressure receptacle (TPR) 4, 266, 318
   legislation 265
Transportable Pressure Equipment Directive 99/36/EC 270
   technical specifications 266
Transverse bend test 208
Transverse tensile test 205
TRD 96, 109, 226
Tube patterns 52
Tubes, steel 369
Tubular Exchanger Manufacturers' Association (TEMA) 47

Ultrasonic testing (UT) 213
   of castings 214
   of welds 216
Unfired pressure vessels 385

ASME V stamp 131
Vacuum leak testing 125
Valves 4
   dimensions 145
   leak testing 150
   materials 145
   technical standards 141
Vessel:
   classes 25
   code applications, NDT 225
   design 21
   dimensional check 116
   distortions 121
   head 25
   stresses 26
   markings 118
   misalignment 121
   openings 90
   'statutory' certification 253
   supports 34
Visual examinations 114, 209
Visual inspection of welds 210

Water tube boilers 107, 361
Water-spray desuperheater 176
Water-tube boilers 108
Waterwall corrosion 243
Weld:
   orientation 196
   preparations, terminology 197
   processes 195
   types 196
Welded joint categories (ASME VII-1) 83
Welded joint efficiency 23
Welding 5
   controlling documentation 204
   procedures 203
   standards 203
   symbols:
      American 200
      BS 198
Weldolet 155
Welds:
   destructive testing of 205
   UT of 216
   visual inspection of 210
Written Scheme of Examination (WSE) 319, 322, 323
Written schemes 319